Attenuated Total Reflectance Spectroscopy of Polymers

Attenuated Total Reflectance Spectroscopy of Polymers

Theory and Practice

Marek W. Urban

American Chemical Society, Washington, DC

Library of Congress Cataloging-in-Publication Data

Urban, Marek W., 1953–
  Attenuated total reflectance spectroscopy of polymers: theory and practice /
Marek W. Urban.

  p. cm.—(Polymer surfaces and interfaces series)
  Includes bibliographical references and index.
  ISBN 0–8412–3348–9
  1. Polymers—Surfaces—Analysis.   2. Reflectance spectroscopy.
  I. Title   II. Series
QD381.9.S97U73 1996
547.7'046—dc20   96–8567
                                                    CIP

The paper used in this publication meets the minimum requirements of American National Standard for Information Sciences—Permanence of Paper for Printed Library Materials, ANSI Z39.48–1984. ∞

About the Author

MAREK W. URBAN is professor and chairman of the Department of Polymers and Coatings at North Dakota State University. He received his B.S. degree in materials science chemistry from the University of Mining and Metallurgy (AGH) in Krakow, Poland, in 1979. His M.S. degree in chemistry was granted from Marquette University in Milwaukee, Wisconsin, in 1981, followed by a Ph.D. degree in chemistry from Michigan Technological University in Houghton, Michigan, in 1984. Before joining North Dakota State University in 1986, he spent two years as a research associate in the Department of Macromolecular Science at Case Western Reserve University in Cleveland, Ohio.

He is the author of more than 200 research papers, numerous review articles, several book chapters, and monographs. His research interests range from surface and interfacial analysis of polymers and coatings to diffusion measurements and molecular level adhesion of polymer networks. For his pioneering work on rheophotoacoustic FTIR spectroscopy, he was awarded the 1990 Megger's Award given by the Federation for Analytical Chemistry and Spectroscopy Societies. He is also credited with patents in this area. For five consecutive years, from 1986 to 1991, he was honored by the 3M Company with the Young Investigator Awards. In 1996 he was awarded the Alcoa Research Foundation Award. He is an invited speaker at many international conferences and Gordon Research Conferences and serves as a lecturer on spectroscopy workshops.

He is the author of *Vibrational Spectroscopy of Molecules and Macromolecules on Surfaces* (John Wiley & Sons) and has edited six ACS Symposium Series and Advances in Chemistry Series books. He is also the editor of the ACS book series Polymer Surfaces and Interfaces, and serves on the advisory boards of numerous professional journals. At North Dakota State

University, in addition to his duties as chairman, he is a director of the National Science Foundation Industry/University Cooperative Research Center in Coatings and the coatings science courses.

Advisory Board

Contents

Foreword

Books published in the Polymer Surfaces and Interfaces Series are designed to provide comprehensive coverage of experimental and theoretical aspects of polymer surfaces and interfaces as well as address practical topics in the field. The author's presentation of the material will reflect the highest scientific standards. Books published in the series will be of interest to polymer scientists and engineers and materials chemists.

MAREK W. URBAN, Editor
Polymer Surfaces and Interfaces Series

Preface

ONE OF THE MOST POPULAR AND USEFUL BRANCHES of surface infrared analysis is referred to as attenuated total reflectance (ATR) spectroscopy. Because of its popularity, there is a need for an overview and a significant update. Although the basic principles of light-attenuated spectroscopy were known at the turn of the 20th century, ATR technique was originated in the early 1960s. For more than three decades, it was used and modified, offering new experimental arrangements and innovative approaches.

In spite of its popularity and attractiveness, one of the main challenges was a lack of reliable quantitative analysis. These studies were initiated by R. N. Jones and continued by others whose contributions stimulated me to further investigate the many gray areas because the proposed approaches were not suitable for simultaneous analysis of strong and weak IR bands. It turned out that one approach was suitable for weak bands, whereas the other allowed ATR analysis of strong bands only. Upon detailed analysis of optical theories and past approaches, it became clear that the double Kramers–Kronig transform (KKT) and the direction of conversion of the refractive index and absorption index spectra make a tremendous difference in quantitative spectral analysis. These studies resulted in the development of several algorithms that allow quantitative analysis of ATR measurements.

This volume attempts to provide a comprehensive theoretical treatment of optical theories relevant to ATR applications in polymer analysis, and it is presented in three parts. The first three chapters are intended for those who are interested in entering the field of surface ATR. This portion focuses on basic principles and experimental features of ATR, without details concerning optical effects and their influence on spectral features. The next section relates to the theoretical considerations that play a key role in interpretation of the ATR spectra. The use of KKT along with limitations and fundamentals of the surface–depth profiling are the primary emphases of this portion. The third section of the book features practical and experimental aspects of ATR, including in situ measurements as well as numerous examples relevant to the use of the technique.

I hope that this volume is of some use, especially for learning modern spectroscopy, which, in order to be used for determining structures of polymer surfaces and interfaces, requires a fundamental understanding of current methods.

Acknowledgments

Numerous papers are published every year on ATR spectroscopy of polymers. I apologize to those whose work on ATR was not included in this volume, but it would be impossible to cover and discuss all of the results published in the literature.

One group of students and postdoctoral researchers contributed significantly to this volume. Special thanks are extended to my current students and postdoctorals: Brian Pennington, Anneke Kaminski, B.-J. Niu, Hueng Kim, Claudia Allison, Lara Tebelius, Joe Stegge, Quien Han, and Yaziu Zhao, who provided valuable comments and help in their special ways to this book. I am particularly thankful to J. B. Huang, a former graduate student in my group, who generated several plots.

This volume would never have been finished in a timely manner if not for the endless efforts of my office staff, Debbie Shasky and her assistant Carrie Cahill, whose tireless efforts deserve a lot of credit.

I would like to particularly recognize Anne Wilson of ACS Books, whose efforts helped me envision the Polymer Surfaces and Interfaces Series, and Barbara Pralle of ACS Books who convincingly stimulated my interest in the series. I am also grateful to Amie Jackowski of ACS Books who has done a magnificent job of converting a manuscript into a book.

Finally, my family deserves recognition, in particular my wife, Dr. Kasia Urban, and my daughter, Ania, who somehow put up with the daily lifestyle of a university professor and writer.

MAREK W. URBAN
North Dakota State University
Fargo, ND 58105

January 1996

Section I

Principles and Basic Concepts

1

Introduction to Basic Concepts

Introduction

Spectroscopy is a scientific discipline concerned with the interactions of electromagnetic radiation with matter. If such radiation impinges on the surface of an object, interactions between the incident beam and the molecules of the object will modify the incident radiation. For example, the incident beam can be reflected, scattered, transmitted, or absorbed, and these modifications carry information about the molecular or physical properties of the object. A spectroscopic experiment allowing detection of reflected, transmitted, scattered, or absorbed light is analogous to the ability of a person to see objects in colors. We need a detector (such as a human eye) capable of distinguishing between colors, an object, and a source of light. Human eyes cannot detect colors if there is no light. Place yourself in a dark room and try to see objects surrounding you. If there is no light, there is no color. The same principle applies to a spectroscopic experiment: there must be a source of light or excitation, an object of study, and a suitable detector.

While an object and a detector are defined by specific chemical and physical properties, light needs further consideration. What is light? Light is an electromagnetic wave. In its simplest, monochromatic form, light can be represented as polarized, oscillating electric and magnetic fields that propagate in space, as depicted in Figure 1.1. The electric and magnetic vectorial components are orthogonal to each other and to the direction of propagation.

When such electromagnetic radiation impinges on a specimen, rays of the incident beam may be reflected, scattered, transmitted, or absorbed. Depending on the experimental arrangements, various rays may be detected, and the total amount of incident energy is the sum of reflected, elastically or inelastically scattered, absorbed, and transmitted light. (A simplified experimental setup is illustrated in Figure 1.2.) This process can be expressed by the following relationship (1):

$$I_0 = I_R + I_A + I_S + I_T \tag{1.1}$$

3348–9/96/0003/$15.00/0/© 1996 American Chemical Society

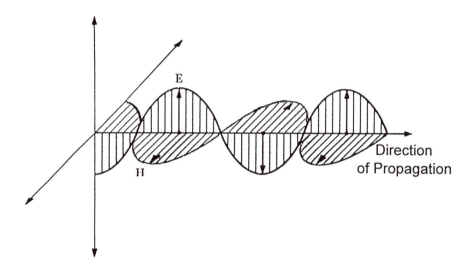

*Figure 1.1. Propagation of a linearly polarized electromagnetic wave in the direction of propagation. Electric (**E**) and magnetic (**H**) vectors are always perpendicular to each other and the direction of propagation. In unpolarized light, the electric component **E** is randomly oriented in an infinite number of directions but remains always perpendicular to the direction of propagation.*

where I_0 is the intensity of the incident beam and I_R, I_A, I_S, and I_T are the intensities of the reflected, absorbed, scattered, and transmitted beams, respectively. The intensity of each beam depends on the intensity and wavelength of the incident radiation, the optical properties of the specimen, the concentrations of species, and the geometry of the experimental setup.

Let us consider what happens to the electromagnetic radiation when a sample is inserted between a source of light and a detector, as in Figure 1.2, and the sample absorbs a fraction of the incident radiation. We will neglect reflection and scattering and measure the amount of light transmitted by the specimen. For that purpose we need to measure the ratio of the sample attenuated (I_0) and nonattenuated (I) intensities, which is proportional to the transmittance of the sample. This relationship is quantitatively related to the chemical composition of the sample by the Beer–Lambert law:

$$I_0/I = e^{-A(\tilde{v})} = e^{-c_2 \varepsilon(\tilde{v})l} \qquad (1.2)$$

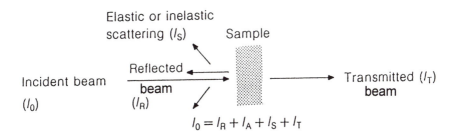

Figure 1.2. Interaction of light with matter. The positioning and sensitivity of the source–detector configuration determines which portion of light will be detected after light impinges on a sample. (Reproduced with permission from reference 1. Copyright 1993 Wiley & Sons, Inc.)

where $A(\tilde{v})$ is the sample absorbance at a given wavenumber $\tilde{v}$, c_2 is the concentration of the absorbing functional group of a sample, $\varepsilon(\tilde{v})$ is the wavenumber-dependent extinction coefficient, and l is the film thickness for the IR beam at a normal incidence to the sample surface. Because the extinction coefficient ε, and therefore the absorbance A, are functions of the wavenumber $\tilde{v}$, quantitative analysis based on the Beer–Lambert law requires a calibration curve. That is, to determine an unknown concentration, a plot of absorbance versus known concentrations should be constructed. Under such circumstances, quantitative analysis using the Beer–Lambert law will be possible if the intensity of the band of interest has absorbance values between 0 and 1 and is usually in a range for which the relationship between absorbance and concentration is linear. This requirement is particularly important in Fourier transform IR spectroscopy because when the absorbance is greater than 1, the commonly used apodization functions* may result in substantial deviations between the measured and true absorbance. Keeping the absorption value between 0 and 1 requires preparing a thin film or dispersing the sample in an IR-transparent medium (for example, a KBr pellet or a mull of mineral oil). Such sampling procedures, however, may lead to major limitations, since they are considered destructive and do not provide surface selectivity.

A simple transmission experiment as described above may provide a significant amount of information concerning molecular structures and entities, and numerous applications have been published in the literature. One apparent drawback is that the technique is not applicable to surface analysis. For that reason, more surface-sensitive sampling techniques have

---

Apodization is a modification of an interferogram in which it is multiplied by a weighting function whose magnitude varies with retardation. Usually, such a function is triangular, quadrupolar, or of a higher order. Further discussion concerning evaluating apodization functions can be found in the literature (*2*).

been developed, including attenuated total reflection (ATR), photoacoustic (PA), reflection–absorption (RA), specular reflectance, emission spectroscopy, and diffuse reflectance infrared Fourier transform (DRIFT). These methods are schematically depicted in Figure 1.3. In ATR spectroscopy (b), light passes through an ATR crystal and interacts with the sample at the interface. In photoacoustic spectroscopy (c), modulated light is absorbed and generates heat, which causes pressure fluctuations in the coupling gas. In RA spectroscopy (d), the IR light incident at an angle between 70° and 89.5° penetrates the sample first and is reflected by highly reflective metal substrates. In specular reflectance spectroscopy (e), the IR light is reflected directly from a reflective polymer surface. In emission spectroscopy (f), the polymer sample on a metal substrate is heated to a high temperature and emits characteristic IR light. In diffuse reflectance spectroscopy (g), the IR light scattered by a polymer sample is collected by a mirror system and directed to the detector.

For all these approaches to surface analysis, several issues are of particular practical interest: (1) whether there are any new chemically distinct structures on the surface in contrast to the bulk of the sample, (2) the concentration of the surface species and how it changes as we look deeper into the surface, and (3) how to conduct nondestructive surface and interfacial measurements. Although the ATR approach requires contact between the sample and an optical element, the simplicity in recording a good-quality spectrum made this technique so popular that almost every analytical laboratory has an arsenal of attachments and crystals. Today's technologies require not only qualitative analysis but also the ability to determine quantities as small as possible. Problems with quantification can arise from using data recorded in ATR and directly applying the Beer–Lambert law, which is valid only in transmission measurements. Because quantification of ATR measurements is important, this book attempts to treat this problem in more detailed fashion. However, before we focus our attention on the quantitative aspects of ATR, let us establish basic principles governing interactions of light with matter. Specifically, reflective and refractive properties of light propagating through media having distinctly different optical properties are of interest because they will dictate not only the amount of energy propagating through but also the direction of propagation.

Principles of Light Reflection and Refraction

When electromagnetic radiation strikes an interface between media of two different refractive indices, refraction and reflection can occur. The law that governs the reflection process requires that the angle of incidence be equal to the angle of reflection (Figure 1.4). In this case, reflection is spec-

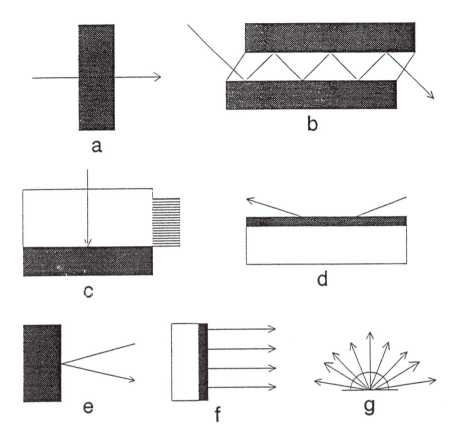

Figure 1.3. Infrared techniques: (A) transmission, (B) attenuated total reflectance, (C) photoacoustic, (D) reflection–absorption, (E) specular reflectance, (F) emission, and (G) diffuse reflectance.

ular. Although the law holds regardless of the surface roughness, it should be remembered that the root-mean-square surface roughness should be much smaller than the wavelength of the incident beam. Otherwise, the angles of incidence will be random, giving reflections in all directions and resulting in diffusely scattered light (Figure 1.5).

    Before we consider the light beam entering an object at various angles and identify how the angle of incidence affects the light path, let us realize that the refractive index of a medium is a measure of its interaction with radiation and is defined by the following relationship:

$$n_v(\lambda) = c/v_v(\lambda) \tag{1.3}$$

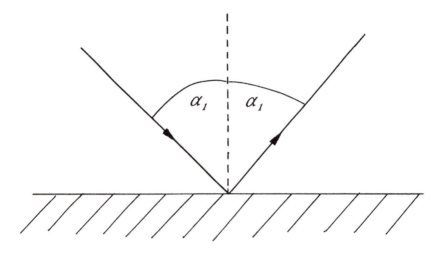

Figure 1.4. Reflection of light propagating through a boundary between media with refractive indices n_1 *and* n_2.

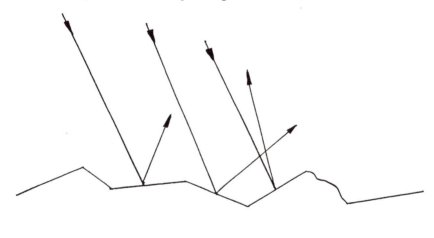

Figure 1.5. Reflection of light from a rough surface. Light is diffusely scattered.

where $n_v(\lambda)$ is the refractive index at a specified wavelength λ,* $v_v(\lambda)$ is the velocity of the radiation in the medium, and c is the velocity in vac-

---

*Electromagnetic radiation is characterized by wavelength λ, frequency v, or wavenumber $\tilde{v}$. Wavenumbers (cm^{-1}) are related to other units by $\tilde{v} = 1/\lambda$, $\tilde{v} = v/c$, where c is the velocity of light (2.997925×10^{10} cm/s).

uum. Because the velocity of radiation is wavelength-dependent, refractive index is also wavelength-dependent. The variation of refractive index with wavelength of radiation is referred to as dispersion. Dispersion curves are complex functions that typically exhibit a gradual increase in refractive index with decreasing wavelength but may show sharp changes. The changes always occur at frequencies that correspond to the harmonic frequency associated with some part of a functional group, molecule, or ion of the substance through which the light travels. At such a frequency, permanent energy transfer from radiation to the substance occurs, and radiation is absorbed.

If electromagnetic radiation passes from one medium to another that has a different physical density, a sudden change of beam direction is detected because of the differences in propagation velocity through two media. If light propagates through a medium with refractive index n_1 and enters a medium with refractive index n_2 (Figure 1.6), the light path will change, and the extent of refraction is given by the following relationship:

$$\frac{\sin\alpha_1}{\sin\alpha_2} = \frac{n_2(\tilde{v})}{n_1(\tilde{v})} = \frac{v_1}{v_2} \tag{1.4}$$

where α_1 and α_2 are the angles of incidence and refraction, respectively, and v_1 and v_2 are propagation velocities in media 1 and 2, respectively. If medium 1 is vacuum, velocity in medium 1 is c, and n_1 is 1. Under such circumstances, n_2 is the ratio of the sines of the angles α_1 and α_2. The principle of refraction, known as Snell's law, is illustrated in Figure 1.6 and is given by eq 1.4.

When electromagnetic radiation strikes an interface between media 1 and 2 that have different refractive indices, reflection also occurs. The fraction of radiation that is reflected becomes larger with increasing differences in refractive index. For a beam that travels normal to the interface, the fraction of radiation reflected is given by

$$R = \frac{I_R}{I_0} = \frac{(n_2(\tilde{v}) - n_1(\tilde{v}))^2}{(n_2(\tilde{v}) + n_1(\tilde{v}))^2} \tag{1.5}$$

where I_0 and I_R are the intensities of incident and reflected radiation, respectively. The ratio R is referred to as *reflectance*. If n_2 is greater than n_1, a 180° phase shift occurs between the incident and reflected waves. As an example, let us consider a zinc selenide crystal, with refractive index n_2 = 2.4, and air, with refractive index n_1 = 1. In this case, the reflectance is equal to 0.17. That is, 17% of the incident power will be reflected.

Snell's law is an important phenomenon in numerous applications involving total internal reflection. Total internal reflection occurs when light traveling in an optically dense medium (one having a high refractive index)

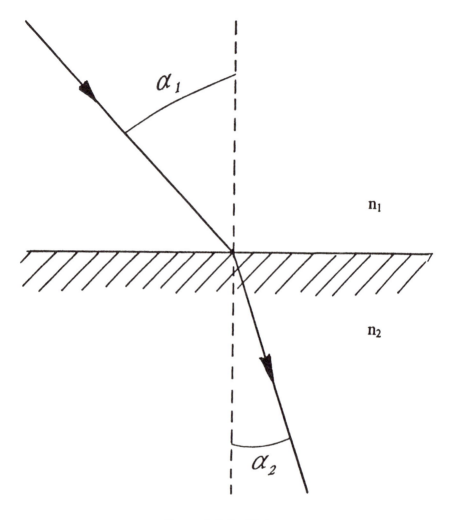

Figure 1.6. Snell's law.

impinges on an interface with a less dense medium. As illustrated in Figure 1.7, when the angle of incidence α_1 equals the critical angle θ_c, $\alpha_2 = 90°$. In this case, all the power is reflected, and Snell's law can be rewritten as

$$\theta_c = \arcsin(n_2(\tilde{v})/n_1(\tilde{v})) \tag{1.6}$$

This relationship forms the basis for fiber optics and total internal reflection spectroscopy.

When the angle of incidence is equal to $\arctan(n_2/n_1)$ and the light is linearly polarized parallel to the plane of incidence, the reflected light

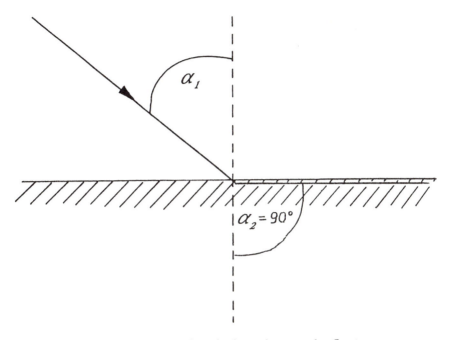

Figure 1.7. Conditions under which total internal reflection occurs.

changes phase and becomes perpendicularly polarized. The angle between the refracted and reflected rays is 90° (Figure 1.8). This observation provides the bases for Brewster windows and laser optics. While further details concerning fundamental principles of optics can be found in the literature (*3, 4*) let us compare transmission and ATR experiments.

Transmission and ATR Experiments

In a simple transmission experiment (Figure 1.3a), the light beam enters the sample normal to its surface and passes through. Let us reposition the light path and shine it on the sample surface at an angle other than 0° to the surface normal. If the sample surface is reflective enough, the sample will reflect most of the incident radiation. This forms the basis for a specular reflectance experiment (Figure 1.3e). However, if an IR-transparent crystal is sandwiched by the sample (Figure 1.3b) and the refractive index of the crystal is greater than that of the sample, then in certain geometric arrangements light will travel by reflecting from the sample–crystal interface. When the light reaches the other end of the crystal, it will carry information about the surface of a sample.

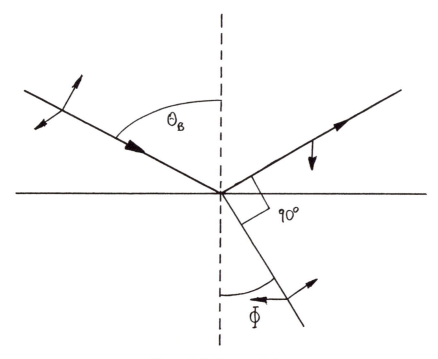

Figure 1.8. Brewster's law.

At this point, let us consider the optical and geometric arrangements necessary for recording ATR spectra. When light strikes the interface of two media with different refractive indices, such as an IR-transparent crystal sandwiched by a sample, a fraction of the light will be reflected, and a fraction may be transmitted, depending on the optical and geometric properties of the experimental setup. If the light initially passes through a medium with a higher refractive index, the angles at which the incoming light can be transmitted or reflected at the interface are depicted in Figure 1.8. In this case, the angle of reflection will be determined by Snell's law, $n_1 \sin \alpha_1 = n_2 \sin \alpha_2$.

When light passes through two media with different refractive indices and the media are in contact with each other, the path of the light will be distorted, depending on the angle of incidence. Figure 1.9 depicts the effect of the angle of incidence on direction of propagation. Light is transmitted at a 90° angle of incidence and partially reflected at $\alpha_1 < \theta_c$ or totally reflected at $\alpha_1 > \theta_c$. When the angle of incidence α_1 is greater than θ_c, the light is totally reflected, and this forms the basis for internal reflection spectroscopy and its extension, ATR spectroscopy.

Transmitted

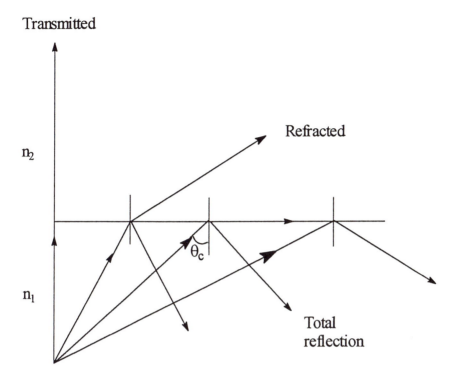

Figure 1.9. Effect of angle of incidence on direction of propagation. (Reproduced with permission from reference 1. Copyright 1993 Wiley & Sons, Inc.)

Another consideration is the amplitude change of the incoming electromagnetic radiation. We need to distinguish between parallel and perpendicular polarized light, because the electric vector may vibrate in the plane perpendicular or parallel to the plane of incidence, so its amplitude will depend on the polarization conditions. These amplitudes can be determined from Fresnel's equations for parallel (also known as transverse electric, or TE) polarized light or for perpendicular (also known as transverse magnetic, or TM) polarized light:

$$r_{\parallel} = \frac{\tan(\alpha_1 - \alpha_2)}{\tan(\alpha_1 + \alpha_2)} \tag{1.7}$$

and

$$r_{\perp} = \frac{\sin(\alpha_1 - \alpha_2)}{\sin(\alpha_1 + \alpha_2)} \tag{1.8}$$

respectively.

The cases shown in Figures 1.4 and 1.6 represent internal reflection. The light strikes the interface from the optically denser medium, which for an ATR experiment is the ATR crystal, and the reflectivities for parallel and perpendicular polarizations can be calculated from Fresnel's equations. A plot of the reflectivity as a function of the angle of incidence for both polarizations is shown in Figure 1.10. The plot indicates that for the parallel and perpendicular polarizations there is one angle of incidence above which 100% reflectivity is achieved. This angle is called the critical angle, θ_c, and the light path for the case where $\alpha_1 < \theta_c$ is shown in Figure 1.4, whereas if $\alpha_2 > \theta_c$, total internal reflection occurs. Therefore, to obtain a distortion-free spectrum, it is necessary during an ATR experiment to stay away from θ_c. The plots in Figure 1.9 were determined for the most commonly used ATR crystals, KRS-5 (thallous bromide iodine) and Ge. They teach us that before selecting the angle of incidence, it is always desirable to assess the critical angle. For KRS-5, with a refractive index of 2.38, assuming that the refractive index for a typical polymer is 1, θ_c will be about 40°.

Thin Film Layers and Multiple Reflections

When light strikes the surface of a thin film, it is decomposed into many components, which in transmission spectroscopy results in so-called interference fringes, seen in a spectrum as a wavy background with the maxima equally spaced. What is the origin of these waves? As illustrated in Figure 1.11, when light passes through the first surface layer (A), a fraction of the incoming beam will be reflected (A_1). A fraction that passes through is again reflected from surface B, and the process continues. When the amplitudes of light reflected from surfaces A and B are the same, the sum of the amplitudes of multiplied reflected components—that is, B—is precisely equal to the amplitude of the first component A_1. However, when these multiple reflected components arrive at the top surface 180° out of phase with respect to A_1 a phase matching occurs, and the rays cancel each other out at the top surface. As a result of this, 100% of the light will be transmitted. In contrast, if there is no phase and amplitude matching, a fraction of the light will be reflected. Whereas the latter can cause many problems in transmission measurements resulting from interference fringes, it is advantageous in an ATR experiment because the first reflected component undergoes a phase change of 180°, followed by no further phase changes for the remaining internal reflections or for transmitted components. It can be shown that in order for the reflectivity to be zero, the optical thickness must be either zero or equal to an integral number of half waves. Under such circumstances, all the multiplied reflected components are in phase with each other but arrive out of phase with respect to A_1. The relationship

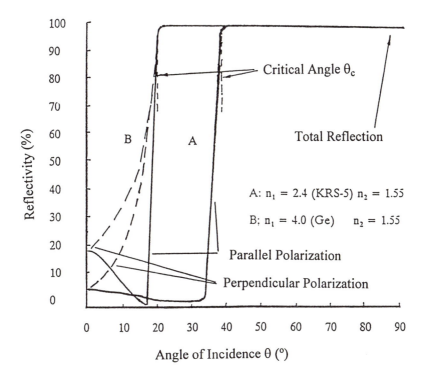

Figure 1.10. Reflectivity plotted as a function of the angle of incidence for internal reflection experiment using KRS-5 (A) and Ge (B) crystals. Solid lines: parallel (TM) polarization. Dashed lines: perpendicular (TE) polarization.

between the amount of light reflected from a thin film and the phase difference is expressed by the following equation:

$$R = \frac{F \sin^2(\gamma/2)}{1 + F \sin^2(\gamma/2)} \tag{1.9}$$

where $F = 4R_1/(1 - R_1)^2$, $R_1 = A_1^2 = B_1(1 - A_1)^2$, $\gamma = 4\pi n_2 x \cos(\alpha_1/\lambda)$, x is the film thickness, n_2 is the refractive index of the film, α_1 is the internal angle of incidence, and λ is the wavelength in vacuum. For $\gamma = 2\pi a$ ($a = 1, 2, 3, ...$), the total reflected light R will be zero. For normal incidence, that is for $\alpha_2 = 0$ ($\cos \alpha_2 = 1$), 100% of the light will be transmitted when $n_2 x = a(\lambda/2)$.

When all the light is internally reflected, the reflectivity is 1. But when a portion of the light is reflected, the reflectivity can be expressed as $R = 1 - \varepsilon_e d_e$, where d_e is an effective thickness and for single reflection $\beta = (100 - R)\%$ is related to the absorption parameter via $d_e = \beta/\varepsilon$. The extinction coef-

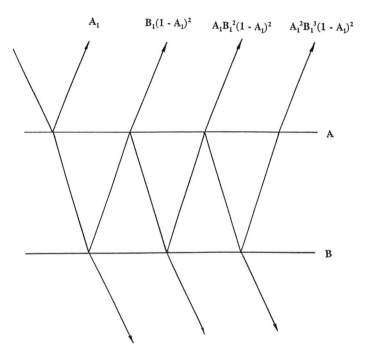

A_1 $B_1(1 - A_1)^2$ $A_1 B_1{}^2(1 - A_1)^2$ $A_1{}^2 B_1{}^3(1 - A_1)^2$

A

B

Figure 1.11. Light components reflected from the surface of a thin surface layer. When $A_1 = B_1(1 - A_1)^2$, the sum $A_1 B_1^2(1 - A_1)^2 + A_1^2 B_1^3 (1 - A_1)^2 + \cdots$ is exactly equal to A_1, and the entire reflected amplitude is equal to zero.

ficient ε_e exhibits more complex dependence than the extinction coefficient in the Beer–Lambert law (eq 1.2). For multiple reflections, the reflectivity will be given as $R^N = (1 - \beta d_e)^N$, where N is the number of reflections. For $\beta d_e \ll 1$, $R^N \cong 1 - N\beta d_e$. Typically, ATR cells are capable of 8 to 20 reflections. For example, for 10 reflections at $45°$ and 4000 cm^{-1}, a penetration depth of 4.15 μm would be expected, whereas at 1000 cm^{-1} the penetration depth would be approximately 16.6 μm. Going down to 600 cm^{-1} will give a depth of around 20 μm.

Experimental conditions for transmission measurements require obeying the Beer–Lambert law, and the absorbance spectrum is expressed in absorbance per micrometer. In other words, for the absorbance spectrum $\beta_2(\tilde{v})$, the band intensities will depend on the absorption and refractive indices of the sample, $k_2(\tilde{v})$ and $n_2(\tilde{v})$, expressed by the following relationships:

$$\beta_2(\tilde{v}) = 4\pi\tilde{v}k_2(\tilde{v}) \times 10^{-4} \tag{1.10}$$

$$k_2(\tilde{v}) = \frac{-2\tilde{v}}{\pi} \int_0^{+\infty} \frac{n_2(s)}{s^2 - \tilde{v}^2} ds \tag{1.11}$$

where ν is the wavenumber (cm^{-1}), and s is the integral variable. In essence, the integral from 0 to $+\infty$ in eq 1.9 represents half of the band of interest. Substituting $k_2(\tilde{\nu})$ from eq 1.8 into eq 1.7 gives the absorbance spectrum $\beta_2(\tilde{\nu})$. In contrast to the transmission measurements, ATR experiments involve reflection of light from the sample–crystal interface. The sample is placed directly on an IR-transparent crystal, and the reflected signal is measured. However, in the reflected signal, the ATR spectrum intensities (which are expressed as a negative logarithm of reflectance) show substantial dispersion around the absorption bands, and their magnitude strongly depends on wavenumber. Therefore, ATR band intensities and positions are distorted, and to compensate for the dispersive relationships around the absorption bands, we need to know the refractive index at the infinite wavenumber ($n_{2,\infty}$), and using these values, calculate the wavenumber-dependent refractive index spectra. At the ATR crystal–sample boundary, the response of the sample to the local evanescent wave can be characterized by a complex refractive index defined as $\hat{n}_j = n_j - ik_j$, where the imaginary part k_j is the absorption index spectrum. In essence, an ATR spectrum is a complex spectrum of the absorption (k) and refractive (n) indices, which are related to each other. While further details of the relationships between k, n, and ATR spectra are disclosed in the remaining chapters, Figure 1.12 illustrates an example of k, n, and ATR spectra, and the appendix provides several examples of ATR configurations and some experimental tips.

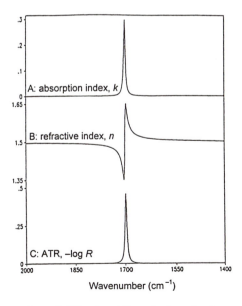

Figure 1.12. Examples of (A) absorption index spectrum, k; (B) refractive index spectrum, n; (C) ATR spectrum, −ln R.

References

1. Urban, M. W. *Vibrational Spectroscopy of Molecules and Macromolecules on Surfaces*; Wiley: New York, 1993.

2. Rabolt, J.; Bellar, R. *Appl. Spectrosc.* **1981**, *35*, 132.

3. Anderson, R. J.; Griffiths, P. R. *Anal. Chem.* **1975**, *47*, 2339.

4. Jenkins, F. A.; White, H. E. *Fundamentals of Optics*, 4th ed.; McGraw–Hill: New York, 1976.

5. Haus, H. A. *Waves and Fields in Optoelectronics*; Prentice–Hall: Englewood Cliffs, NJ, 1984.

6. Harrick, N. J. *Phys. Rev. Lett.* **1960**, *4*(5), 224.

7. Harrick, N. J. *Phys. Chem.* **1960**, *64*, 1110.

8. Fahrenfort, J. *Spectrochim. Acta* **1961**, *17*, 698.

2

Attenuated Total Reflectance Spectroscopy as a Surface Tool

Introduction

While measurements conducted in transmission mode have certain limitations related to the lack of surface sensitivity, in numerous applications (for example, degradation studies, surface reactions and modifications, and forensic analysis) it is essential that the information coming from the bulk of the material be distinguished from that coming from the surface or interface. For that reason transmission spectroscopy may fail because, in addition to the fact that such measurements often require destructive sample preparations, the technique is not surface-sensitive; and surface sensitivity is where ATR spectroscopy surpasses other analytical methods. It is a convenient, surface-sensitive analytical tool. All the analysis requires is aligning an ATR attachment, pressing the sample against an ATR crystal, and collecting spectra. It should be realized—and this is demonstrated throughout this text—that ATR data have often been subjected to Beer–Lambert law analysis, neglecting the fact that ATR spectroscopy is a contact method and that optical effects from the contact between a sample and a crystal may result in IR band distortions and other problems. This book outlines methods that allow use of the Beer–Lambert law in ATR spectroscopy to make the measurements quantitative and eliminate the influence of crystal–sample contact and crystal coverage area. But before these issues are addressed, let us try to understand why ATR is surface-sensitive and what problems may be associated with the technique.

Why ATR Is Surface-Sensitive

To establish surface sensitivity and to further understand the formalism and theory behind an ATR experiment, let us consider several situations in which light passes through a boundary between two layers with different refractive properties. Such cases were illustrated in Chapter 1. Let us consider once more the crystal–sample interface. This time, however, let us consider a scenario in which light is polarized near the crystal–sample interface. This can be schematically illustrated by considering a reflection model depicted in Figure 2.1, where a plane wave, polarized along the y axis, propagates in the x–z plane at an angle of α_1 from the z axis and, when it impinges on the boundary in the x–y plane, splits into reflected and transmitted beams. Although in reflection spectroscopy only the reflected beam is detected, the observed spectral features are caused by the interaction of the transmitted wave with a sample surface. The electric field of the transmitted wave may be expressed as

$$E_y = E_{y_0}\exp\left(\frac{i2\pi}{\lambda}z\xi_2\right)\exp\left(\frac{i2\pi}{\lambda}y\zeta_2\right)\exp(-i2\pi\nu t) \tag{2.1}$$

where λ is the wavelength in vacuum, ν is the frequency of the light, E_{y_0} is the amplitude of the electric field at the interface, and ξ_2 and ζ_2 are the z and x components of the propagation vector of the transmitted wave (*1*). According to Snell's law, introduced in Chapter 1, these quantities are related by the following equations:

$$\zeta_2 = n_2\sin\alpha_2 = n_1\sin\alpha_1 \tag{2.2}$$

and

$$\xi_2 = n_2\cos\alpha_2 = n_2(1 - \sin^2\alpha_2)^{1/2} = (n_2^2 - n_1^2\sin^2\alpha_1)^{1/2} \tag{2.3}$$

where n_1 and n_2 are the refractive indices of the ATR crystal and the sample, respectively, and α_2 is the angle between the transmitted light and the normal of the interface. It can be seen that when the incident angle α_1 is larger than a critical angle given by $\theta_c = \arcsin(n_2/n_1)$, ξ_2 contains an imaginary component even when the sample is nonabsorbing and therefore n_2 is real. This imaginary component of ξ_2 will result in an exponential decay of E_y as a function of z, and the transmitted wave is called the evanescent wave. Using eq 2.1, we can examine the light energy carried by the evanescent waves in the sample.

When n_2 is real and ξ_2 is imaginary, there is a 90° phase difference between E_y and the x component of the magnetic field, H_x. This results from the following relationship (*1*):

$$H_x = \frac{\lambda}{i2\pi}\frac{\partial E_y}{\partial z} = i\,\mathrm{Im}(\xi_2)E_y \tag{2.4}$$

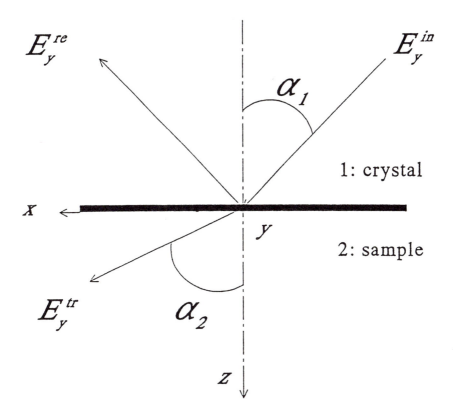

Figure 2.1. A schematic diagram of the perpendicularly polarized light beam (E_y^{in}) impinging on the interface between two semi-infinite media and split into two beams: E_y^{tr}, which is transmitted, and E_y^{re}, which is reflected.

This 90° phase difference will result in a zero time average of the energy propagation along the z axis, which can be calculated according to Poynting's theorem (*1*):

$$\langle s_z \rangle = v \int_0^{1/v} \mathrm{Re}(E_y)\mathrm{Re}(H_x)dt$$

$$= v|E_y|^2 \int_0^{1/v} \cos(2\pi v t)\cos(2\pi v t + \pi/2)dt = 0 \qquad (2.5)$$

This equation shows that any instantaneous propagation of light energy into the sample will come back to the crystal within one reflection.

When the sample absorbs light and n_2 is replaced by a complex number $n_2 - ik_2$, where k_2 is the absorption index, the propagation vector ξ_2

contains a real component, resulting in a nonzero time average for the energy propagation of the evanescent waves along the z axis,

$$\langle s_z \rangle = \frac{1}{2}|E_y|^2 \text{Re}(\xi_2) \tag{2.6}$$

The divergence of this energy propagation represents the energy absorbed per unit thickness,

$$-\frac{\partial}{\partial z}\langle s_z \rangle = \frac{2\pi}{\lambda}|E_y|^2 \text{Re}(\xi_2)\text{Im}(\xi_2)$$

$$= \frac{2\pi}{\lambda}k_2 n_2 |E_{y_0}|^2 \exp\left(-\frac{2z}{d_p}\right) \tag{2.7}$$

where the mathematical identity between $k_2 n_2$ and $\text{Re}(\xi_2)\text{Im}(\xi_2)$ has been used, and the penetration depth d_p is given by

$$d_p = \frac{\lambda}{2\pi \, \text{Im}(\xi_2)} \tag{2.8}$$

According to eq 2.7, the contribution of a species at a given depth to the observed spectrum is an exponential decay function of z, making ATR spectroscopy a surface-sensitive technique.

With the appropriate plane-wave expressions for the incident and reflected light shown in Figure 2.1, and the boundary conditions imposed by the continuities of E_y and H_x, one can derive Fresnel's equation by taking the ratio of E_y^{re} and E_y^{in} :

$$\hat{r} \equiv \frac{E_y^{\text{re}}}{E_y^{\text{in}}} = R^{1/2}e^{i\theta} = \frac{\xi_1 - \xi_2}{\xi_1 + \xi_2} \tag{2.9}$$

where ξ_1 is the z component of the propagation vector of the incident light, θ is the phase difference between the electric fields of the incident and reflected light beams, and R is the reflectivity measured in an ATR experiment. Since R is the ratio of the reflected and incident light intensities, and $1 - R$ is the ratio of the energy flux across the boundary (given by eq 2.6) to the energy flux of the incident light (given by $\xi_1|E_y^{\text{in}}|^2/2$), $1 - R$ can be written as

$$1 - R = \frac{\text{Re}(\xi_2)|1 + r|^2}{\xi_1} = \frac{4k_2 n_2 \xi_1}{\text{Im}(\xi_2)|\xi_1 + \xi_2|^2} \tag{2.10}$$

If the intensities of absorption bands are so low that $k_2 > n_2$ and $n_2 k_2 \ll n_2^2 - k_2^2 - n_1^2 \sin^2\alpha_1$, ξ_2 can be approximated as follows:

$$\xi_2 = \sqrt{n_2^2 - k_2^2 - n_1^2 \sin^2\alpha_1 + ik_2 n_2}$$

$$\approx \sqrt{n_2^2 - n_1^2 \sin^2\alpha_1} \tag{2.11}$$

$$= i \, \text{Im}(\xi_2)$$

and

$$\left|\xi_1 + \xi_2\right|^2 \approx \xi_1^2 + \text{Im}(\xi_2)^2 = n_1^2 - n_2^2 \tag{2.12}$$

Substituting these approximate relationships into eq 2.6, and using the relationship between the absorption index spectrum and the wavelength of light $k_2 = c_2\varepsilon\lambda/4\pi$, where c_2 is the concentration of the absorbing species in the sample and ε is the extinction coefficient, the following equation can be derived:

$$-\ln R \approx 1 - R \approx c_2\varepsilon \, \frac{\lambda n_2 n_1 \cos\alpha_1}{\pi(n_1^2 - n_2^2)\sqrt{n_1^2\sin^2\alpha_1 - n_2^2}} = c_2\varepsilon d_e \tag{2.13}$$

where d_e is the effective thickness, which is the film thickness of the sample that would give the same absorbance in a transmission experiment with the beam at normal incidence as that obtained in ATR. Equation 2.13 accounts for the quantitative principle of ATR spectroscopy and can easily be extended to parallel light polarization. In other words, this is a counterpart of the Beer–Lambert law (eq 1.2) used in transmission experiments. Chapter 3 will provide further necessary derivations and details related to perpendicular and parallel polarized light. Because eq 2.13 for $-\ln R$ is a counterpart to the absorbance spectrum recorded in a transmission mode of detection, this relationship allows ATR measurements to be treated as a technique for recording absorbance spectra of surfaces, and the equation will serve for quantitative analysis.

Problems with ATR

In addition to the surface-sensitivity, the following spectral features that distinguish ATR spectroscopy from other techniques have been correctly recognized in the literature: (1) long-range frequency effect on sensitivity caused by the 1 factor contained in the expression for d_e (eq 2.13) and (2) frequency shifts of the band maximum caused by dispersion of the refractive index spectra across the absorption band. The first feature suggests that when normalizing ATR spectra, the normalization band should be relatively close to the band of interest because the depth of penetration is wavenumber-dependent (2). However, if such bands do overlap, one would expect band distortions and intensity changes, and these effects will be described quantitatively in Chapter 7. The second factor is also important when interpreting the frequency shifts of the surface species relative to the bulk. If these two features were the only difference between ATR and true absorbance spectra, many data-processing techniques originally developed for transmission measurements, such as spectral subtraction, the ratio method (3), normalization against an internal reference band (4), and multivariate statistical methods (5, 6), would remain valid with minimal adjustment. An additional distinctive feature of ATR spectroscopy that has often been neglected is the interferences between overlapping absorption bands. In Chapter 4 the magnitude of such interferences will be illustrated as a function of band intensity and band separation, and the discussion will focus on the consequences of ignoring these interferences.

Experimental methods employed in an ATR experiment may also lead to artificial effects. In this category, frequency shifts and changes of relative band intensities are the main features. Such artificial effects may be eliminated by converting the reflectivity spectrum into the optical constants of the sample, which can in turn be used to calculate absorbance spectra. While the early approaches to such conversions were based on the analysis of two ATR spectra of the same sample measured at different experimental conditions (7), more convenient methods based on the Kramers–Kronig relations were developed (8, 9), and these will be discussed later. The KKT-based algorithms will be assessed in terms of their precision and accuracy. One of the major advantages of these approaches is that the extinction coefficients can be calculated without sample thickness measurements (10).

As indicated earlier, surface sensitivity is one of the most acclaimed features of ATR spectroscopy (11–14). However, the assessment of concentrations at specific surface depths recorded at various penetration depths requires complicated theoretical considerations. Previous approaches (15–17), such as inverse Laplace transformation, are based on weak band assumptions and do not consider the effects of the refractive index dispersion across absorption bands. Such effects, however, will result in significant variations in the penetration depths and distortions of the observed spectra, particularly in overlapping regions. The difficulty of considering these effects for nonhomogeneous samples investigated in depth-profiling studies arises from the fact that the Fresnel reflectivity equation has a limited use because eq 2.13 is based on the assumption that the optical constants (n_2 and k_2) are the same throughout the sample. In Chapter 6, the theoretical models will be tested on several practical examples by analyzing low-concentration species with strong overlapping absorption bands. In Chapter 7, polymer surfaces with concentration gradients will be treated as a stack of thin parallel homogeneous layers, which will be analyzed using a matrix theory (18) to establish a correlation between concentrations at various depths and ATR spectra measured at various penetration depths. The depth-profiling theory will be demonstrated by analyzing the surface concentrations of polymeric films.

References

1. Born, M.; Wolf, E. *Principles of Optics*, 5th ed.; Pergamon: Oxford, 1975.

2. Mirabella, J. F. M. *Spectroscopy* **1990**, *5*, 21.

3. Koenig, J. L.; Esposito, L.; Antoon, M. K. *Appl. Spectrosc.* **1977**, *31*, 292.

4. Gillette, P. C.; Lando, J. B.; Koenig, J. L. In *Fourier Transform Infrared Spectroscopy: Applications to Chemical Systems*; Ferraro, J. R.; Basile, L. J., Eds.; Academic: Orlando, FL, 1985; Chapter 1.

5. Vigerust, B.; Kolset, K.; Nordenson, S. *Appl. Spectrosc.* **1991**, *45*, 173.

6. Toft, J.; Kvalheim, O. M.; Karstang, T. V.; Christy, A. A.; Kleveland, K.; Henriksen, A. *Appl. Spectrosc.* **1992**, *46*, 1002.

7. Crawford, B., Jr.; Goplen, T. G.; Swanson, D. In *Advances in Infrared and Raman Spectroscopy*; Clark, R. J. H.; Hester, R. E., Eds.; Heyden: London, 1980; Vol. 4, Chapter 2.

8. Dignam, M. J.; Mamiche-Afara, S. *Spectrochim. Acta* **1988**, *44A*, 1435.

9. Bertie, J. E.; Eysel, H. H. *Appl. Spectrosc.* **1985**, *39*, 392.

10. Bardwell, J. A.; Dignam, M. J. *Anal. Chim. Acta* **1986**, *181*, 253.

11. Blackwell, C. S.; Degen, P. J.; Osterholtz, F. D. *Appl. Spectrosc.* **1978**, *32*, 480.

12. Carlsson, D. J.; Wiles, D. M. *Macromolecules* **1972**, *4*, 173.

13. Webb, J. R. *J. Polym. Sci. Polym. Chem. Ed.* **1972**, *10*, 2335.

14. Hirayama, T.; Urban, M. W. *Prog. Org. Coat.* **1992**, *20*, 81.

15. Tompkins, H. G. *Appl. Spectrosc.* **1974**, *28*, 335.

16. Hirschfeld, T. *Appl. Spectrosc.* **1977**, *31*, 289.

17. Stuchebryukov, S. D. *Surf. Interface Anal.* **1984**, *6*(1), 29.

18. Hansen, W. *J. Opt. Soc. Am.* **1968**, *58*(3), 380.

3

Useful Relationships in ATR Spectroscopy

The history of ATR, which belongs to the internal reflection spectroscopy methods, began more than two centuries ago with Newton's observation that when a propagating wave of radiation undergoes total internal reflection at the interface between two media of different refractive indices, an evanescent field penetrates the interface into the medium with the lower refractive index. However, the exploitation of this phenomenon for generating useful absorption spectra did not begin until the early 1960s, with the pioneering efforts of Harrick (*1*) and Fahrenfort (*2*), followed by experimental work by Sharpe (*3*). Over the past 30 years, numerous applications of ATR spectroscopy have been reported (*4–8*) and reviewed (*9–11*), including practical aspects of ATR experimental setups (*12*). Although ATR spectroscopy normally uses multiple reflections, fundamental relationships can be illustrated by considering an ideal single reflection. This chapter focuses on these relationships, along with their derivations pertinent to a single reflection. Derivations are particularly important because they help in understanding not only the validity, but also the limitations of the commonly used relationships, including those related to Fresnel's equations, depth of penetration, and especially effective thickness.

Reflectance

Let us go back to the reflection model illustrated in Figure 2.1, in which a plane-wave light beam in the ATR crystal propagates in the x–z plane at an angle of α_1 from the z axis. When it impinges on the boundary in the x–y plane, the incident light beam is partially reflected into the ATR crystal and partially transmitted into the sample, whose response to the electromagnetic fields of the light can be characterized by a complex refractive index $\hat{n}_2 = n_2 - ik_2$, where the imaginary part k_2 is the absorption index. This

3348–9/96/0027/$15.25/0/© 1996 American Chemical Society

relationship is illustrated in Figure 3.1, which shows the k_2 and n_2 spectra of two generated bands A and B using Lorentzian shape. In analogy to the transmittance in transmission spectroscopy, the reflectance in reflection spectroscopy may be defined as the ratio between the intensity of the reflected light and that of the incident light:

$$R = \frac{I^{re}}{I^{in}} \qquad (3.1)$$

According to the electromagnetic theory, the light intensity is interpreted as the average amount of electromagnetic energy at a given wavenumber flowing through a unit area in a unit time. The unit area is usually normal to the direction of propagation of the electromagnetic waves. Multiplying the numerator and denominator of eq 1 by $\cos \alpha_1$, we obtain

$$R = \frac{I^{re}\cos\alpha_1}{I^{in}\cos\alpha_1} = \frac{\langle s_z^{re}\rangle}{\langle s_z^{in}\rangle} \qquad (3.2)$$

where $\langle s_z^{re}\rangle$ and $\langle s_z^{in}\rangle$ represent the time-average energy flux through a unit area parallel to the sample–crystal interface (i.e., along the z axis). Such energy fluxes can be calculated using Poynting's theorem,[*]

$$\langle s_z\rangle = \frac{c v}{4\pi} \int_0^{1/\nu} E_{xy}H_{xy}\, dt \qquad (3.3)$$

where E_{xy} and H_{xy} are the projections of the electric and magnetic fields in the x–y plane, and c is the speed of light in vacuum. For simplicity, we consider only two linearly polarized light beams: s- and p-polarized. In s-polarization, also referred to as perpendicular polarization or transverse electric (TE) wave, the electric field is perpendicular to the plane of incidence, and therefore the z and x components, E_z and E_x, are zero and $E_{xy} = E_y$. Because the magnetic field is perpendicular to the electric field, we also have $H_{xy} = H_x$. In p-polarization, also referred to as parallel polarization or transverse magnetic (TM) wave, the magnetic field is perpendicular to the plane of incidence, and therefore H_z and H_x are zero and $H_{xy} = H_y$. Because of the perpendicular relationship between electric and magnetic fields, $E_{xy} = E_x$. Having defined the coordinate system, let us analyze the reflectance for s- and p-polarizations.

---

[*]Because a traveling electromagnetic wave can transport energy from place to place, the rate of energy flow per unit of a cross section in an electromagnetic wave is described by vector $\langle \mathbf{s}\rangle$, known as Poynting's vector. The vectors $\mathbf{E}$ and $\mathbf{H}$ refer to their instantaneous values at a given point, and $\langle \mathbf{s}\rangle$ is the direction of propagation.

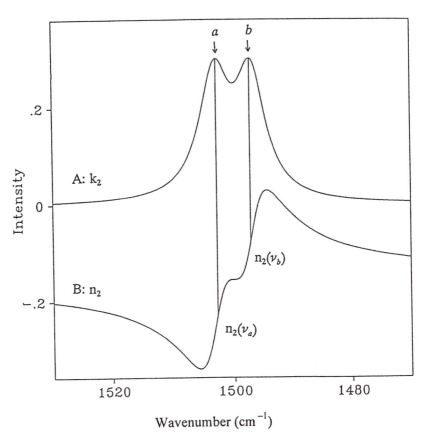

Figure 3.1. Simulated refractive and absorption index spectra for two bands A and B near 1500 cm^{-1}.

Reflectance with s-Polarization

When incident light is s-polarized, the incident electric field can be expressed as

$$E_y^{in} = U^{in} e^{\frac{i2\pi}{\lambda} z \xi_1} e^{\frac{i2\pi}{\lambda} y \zeta_1} e^{-2\pi \nu t} \qquad z \leq 0 \qquad (3.4)$$

where λ is the wavelength in vacuum, ν is the frequency of the light, U^{in} is the amplitude of the electric field of the incident wave at the interface, and ξ_1 and ζ_1 are the z and x components of the propagation vector of the incident wave and are given by $n_1 \cos \alpha_1$ and $n_1 \sin \alpha_1$, respectively.

Similarly, for the electric fields of the reflected and transmitted waves, we can write that

$$E_y^{re} = U^{re} e^{-\frac{i2\pi}{\lambda} z \xi_1} e^{\frac{i2\pi}{\lambda} y \zeta_1} e^{-i2\pi \nu t} \qquad z \leq 0 \qquad (3.5)$$

and

$$E_y^{tr} = U^{tr} e^{\frac{i2\pi}{\lambda} z \xi_2} e^{\frac{i2\pi}{\lambda} y \zeta_2} e^{-i2\pi \nu t} \qquad z \geq 0 \qquad (3.6)$$

where ξ_2 and ζ_2, the z and x components of the propagation vector of the transmitted wave, are given by the previously introduced Snell's law,

$$\zeta_2 = \hat{n}_2 \sin\alpha_2 = n_1 \sin\alpha_1 \qquad (3.7)$$

and

$$\xi_2 = \hat{n}_2 \cos\alpha_2 = \hat{n}_2 (1 - \sin^2\alpha_2)^{1/2} = (\hat{n}_2^2 - n_1^2 \sin^2\alpha_1)^{1/2} \qquad (3.8)$$

where α_2 is the angle of the transmitted light measured from the z axis.

According to eq 3.3, in order to calculate the energy flux along the z axis, it is necessary to derive H_{xy}. As mentioned earlier, for s-polarization $H_{xy} = H_x$. Applying the relationship

$$H_x = \frac{\lambda}{i2\pi} \frac{\partial E_y}{\partial Z} \qquad (3.9)$$

we obtain the following relationships for the magnetic fields of the incident, reflected, and transmitted light beams:

$$H_x^{in} = \xi_1 E_y^{in} \qquad z \leq 0 \qquad (3.10)$$

$$H_x^{re} = -\xi_1 E_y^{re} \qquad z \leq 0 \qquad (3.11)$$

and

$$H_x^{tr} = \xi_2 E_y^{tr} \qquad z \geq 0 \qquad (3.12)$$

The continuity of H_x and E_y at $z = 0$ dictates the following relations:

$$U^{in} + U^{re} = U^{tr} \qquad (3.13)$$

and

$$\xi_1 U^{in} - \xi_1 U^{re} = \xi_2 U^{tr} \qquad (3.14)$$

Combining eqs 3.13 and 3.14, we obtain Fresnel's equation, although in a slightly different form:

$$\hat{r} \equiv \frac{U^{re}}{U^{in}} = \frac{\xi_1 - \xi_2}{\xi_1 + \xi_2}$$

$$= \frac{n_1 \cos\alpha_1 - (\hat{n}_2^2 - n_1^2 \sin^2\alpha_1)^{1/2}}{n_1 \cos\alpha_1 + (\hat{n}_2^2 - n_1^2 \sin^2\alpha_1)^{1/2}} \tag{3.15}$$

The expressions for the electric and magnetic fields allow us to compute the energy flux by using Poynting's theorem, given as follows:

$$\langle s_z \rangle = \frac{cv}{4\pi} \int_0^{1/v} \mathrm{Re}(E_y)\mathrm{Re}(H_x)dt$$

$$= \frac{cv}{4\pi} \int_0^{1/v} \mathrm{Re}(E_y)[\mathrm{Re}(\xi)\mathrm{Re}(E_y) - \mathrm{Im}(\xi)\mathrm{Im}(E_y)]dt$$

$$= \frac{cv}{4\pi}|E_y|^2 \int_0^{1/v} [\mathrm{Re}(\xi)\cos^2(2\pi vt) + \mathrm{Im}(\xi)\cos(2\pi vt)\sin(2\pi vt)]dt \tag{3.16}$$

$$= \frac{c}{8\pi}|E_y|^2\mathrm{Re}(\xi)$$

Using this relationship, the reflectance R can be related to the Fresnel reflectivity $\hat{r}$ through the following equation:

$$R_\perp = \frac{\langle s_z^{re} \rangle}{\langle s_z^{in} \rangle} = \frac{|E_y^{re}|^2}{|E_y^{in}|^2}$$

$$= \frac{|U^{re}|^2}{|U^{in}|^2} = |\hat{r}_\perp|^2 \tag{3.17}$$

$$= \left| \frac{n_1 \cos\alpha_1 - (\hat{n}_2^2 - n_1^2 \sin^2\alpha_1)^{1/2}}{n_1 \cos\alpha_1 + (\hat{n}_2^2 - n_1^2 \sin^2\alpha_1)^{1/2}} \right|^2$$

Reflectance with p-Polarization

When light is p-polarized, the magnetic fields of the three light waves can be expressed as

$$H_y^{in} = V^{in}e^{\frac{i2\pi}{\lambda}z\xi_1}e^{\frac{i2\pi}{\lambda}y\zeta_1}e^{-i2\pi vt} \qquad z \leq 0 \tag{3.18}$$

$$E_y^{re} = V^{re} e^{-\frac{i2\pi}{\lambda}z\xi_1} e^{\frac{i2\pi}{\lambda}y\zeta_1} e^{-i2\pi vt} \qquad z \leq 0 \qquad (3.19)$$

and

$$E_y^{tr} = V^{tr} e^{\frac{i2\pi}{\lambda}z\xi_2} e^{\frac{i2\pi}{\lambda}y\zeta_2} e^{-i2\pi vt} \qquad z \geq 0 \qquad (3.20)$$

where V^{in}, V^{re}, and V^{tr} are the amplitudes of the magnetic fields of the incident, reflected, and transmitted waves, respectively.

Again, according to eq 3.3, in order to calculate the energy flux along the z axis, it is necessary to derive an expression for E_{xy}. Because for p-polarization $E_{xy} = E_x$ it is only necessary to determine E_x, according to the following relation:

$$E_x = \frac{\lambda}{i2\pi n^2} \frac{\partial H_y}{\partial Z} \qquad (3.21)$$

Applying this relation to the electric fields of the three light beams, we obtain again

$$E_x^{in} = \frac{\xi_1}{n_1^2} H_y^{in} \qquad z \leq 0 \qquad (3.22)$$

$$E_x^{re} = -\frac{\xi_1}{n_1^2} H_y^{re} \qquad z \leq 0 \qquad (3.23)$$

and

$$E_x^{tr} = \frac{\xi_2}{\hat{n}_2^2} H_y^{tr} \qquad z \geq 0 \qquad (3.24)$$

The continuity of H_y and E_x at $z = 0$ dictates the following relationships:

$$V^{in} + V^{re} = V^{tr} \qquad (3.25)$$

and

$$\frac{\xi_1}{n_1^2} V^{in} - \frac{\xi_1}{n_1^2} V^{re} = \frac{\xi_2}{\hat{n}_2^2} V^{tr} \qquad (3.26)$$

Combining eqs 3.25 and 3.26, we obtain Fresnel's equation for p-polarized light:

$$\hat{r}_1 \equiv \frac{V^{re}}{V^{in}} = \frac{(\xi_1/n_1^2) - (\xi_2/\hat{n}_2^2)}{(\xi_1/n_1^2) + (\xi_2/\hat{n}_2^2)}$$

$$= \frac{\hat{n}_2 \cos\alpha_1 - n_1(1 - (n_1/\hat{n}_2)^2 \sin^2\alpha_1)^{1/2}}{\hat{n}_2 \cos\alpha_1 + n_1(1 - (n_1/\hat{n}_2)^2 \sin^2\alpha_1)^{1/2}}$$

(3.27)

Again using Poynting's theorem on the expressions for electric and magnetic fields, we can determine the energy flux:

$$\langle s_z \rangle = \frac{cv}{4\pi} \int_0^{1/v} \text{Re}(H_y)\text{Re}(E_x) dt$$

$$= \frac{cv}{4\pi} \int_0^{1/v} \text{Re}(H_y)[\text{Re}(\xi/n_2^2)\text{Re}(H_y) - \text{Im}(\xi/n_2^2)\text{Im}(H_y)] dt$$

$$= \frac{cv}{4\pi}|H_y|^2 \int_0^{1/v} [\text{Re}(\xi/n_2^2)\cos^2(2\pi vt)$$

$$+ \text{Im}(\xi/n_2^2)\cos(2\pi vt)\sin(2\pi vt)] dt$$

(3.28)

$$= \frac{c}{8\pi}|H_y|^2 \text{Re}(\xi/n_2^2)$$

Using this relationship, we can relate the reflectance R to the Fresnel reflectivity $\hat{r}$:

$$R_{\parallel} = \frac{\langle s^{re} \rangle}{\langle s^{in} \rangle} = \frac{|H_y^{re}|^2}{|H_y^{in}|^2}$$

$$= \frac{|V^{re}|^2}{|V^{in}|^2} = |\hat{r}_{\parallel}|^2$$

(3.29)

$$= \left| \frac{n_2 \cos\alpha_1 - n_1(1 - (n_1/n_2)^2 \sin^2\alpha_1)^{1/2}}{n_2 \cos\alpha_1 + n_1(1 - (n_1/n_2)^2 \sin^2\alpha_1)^{1/2}} \right|^2$$

A comparison of eqs 3.17 and 3.29 clearly indicates that to calculate reflectance, we must take into account not only the refractive indices of the crystal and the sample but also the polarization of the light. As a matter of fact, it is no overstatement to say that for quantitative purposes only polarized spectra should be used. Such an approach will have far more important consequences if one is interested in experimentally determining the orientation of the surface species.

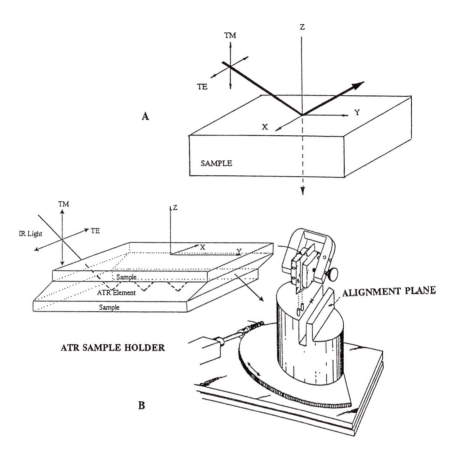

Figure 3.2. (A) Geometry and definition of spatial axis for ATR three-dimensional measurements; (B) ATR sample holder for orientation studies (adapted with permission from ref. 4).

Polarization Experiment

In an effort to characterize surface molecular orientation in three dimensions, Sung (*13, 14*) developed an ATR attachment allowing such measurements. Figure 3.2A illustrates a definition of two polarizations with respect to the sample surface, and Figure 3.2B shows the sample holder. The IR beam enters the crystal from the back face and exits from face B before entering the detector. The entire attachment is mounted on a movable plate so that the incident angle can be adjusted. For depth-profiling studies, 10 angles, ranging from 39.5° to 55.2°, should normally be examined, to provide information from 0.5 to 1.5 μm at 1000 cm^{-1} for a typical polymer film.

For three-dimensional analyses, the polarizer, through which the light passes before it enters the ATR crystal, must be rotated by 90° from the TE to the TM position.

Critical Angle of Incidence and Penetration Depth

Although one would rightly expect that in reflection spectroscopy only reflected light beams are detected, the features of the reflection spectra result from interactions of the transmitted light. Therefore, it is useful to examine the expressions for the transmitted light (E_y^{tr}) given by eqs 3.6 and 3.20. It appears that the intensities of the electromagnetic fields transmitted into the sample will decrease with z if ξ_2 contains a nonzero imaginary part. This situation will occur when the sample absorbs radiation. On the other hand, for a transparent sample ($k_2 = 0$), ξ_2 is imaginary if the angle of incidence α_1 is larger than a critical angle given by

$$\theta_c = \arcsin(n_2/n_1) \qquad (3.30)$$

It is useful to remember that this relationship provides a means for choosing the crystal appropriate for a given sampling situation. For example, for the two most commonly used crystals—KRS-5 and Ge—the relationship between the critical angle of incidence and the refractive index of the sample is shown in Figure 3.3.

The transmitted wave is referred to as an evanescent wave, and according to eqs 3.17 and 3.29, when the real part of ξ_2 vanishes, the reflectance becomes unity. For this reason, such a reflection is referred to as a total reflection. Analysis of eqs 3.16 and 3.28 also indicates that the average energy flux carried by the evanescent wave is equal to zero, and therefore any instantaneous energy flow into the sample will come out within one cycle.

When the incident angle is larger than the critical angle and the sample absorbs radiation (i.e., $k_2 \neq 0$), the real part of ξ_2 is no longer zero, and the reflectance becomes less than unity. Under such circumstances, the amount of electromagnetic energy being put into the sample by the evanescent wave exceeds the amount coming out, and the evanescent wave will be attenuated. The light energy absorbed per unit volume at a depth z can be determined by calculating the negative divergence of the energy flux of the evanescent wave. From eqs 3.16 and 3.28, we obtain

$$-\frac{\partial \langle s_{z,\perp}^{tr} \rangle}{\partial z} = \frac{c\,\mathrm{Re}(\xi_2)\mathrm{Im}(\xi_2)}{2\lambda} \left| U^{tr} \right|^2 e^{-\frac{4\pi}{\lambda}\mathrm{Im}(\xi_2)z} \qquad (3.31)$$

and

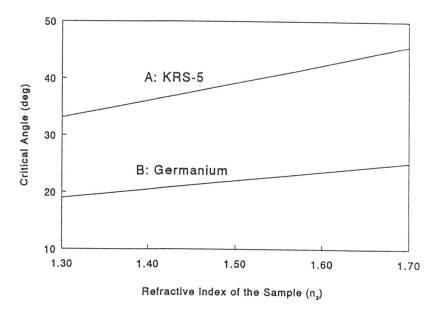

Figure 3.3. Relationship between critical angle and refractive index of a sample for (A) KRS-5 crystal and (B) Ge crystal.

$$-\frac{\partial\langle s_{z,\parallel}^{tr}\rangle}{\partial Z} = \frac{c\mathrm{Re}(\xi_2/\hat{n}_2)\mathrm{Im}(\xi_2)}{2\lambda}\left|V^{tr}\right|^2 e^{-\frac{4\pi}{\lambda}\mathrm{Im}(\xi_2)z} \qquad (3.32)$$

According to these equations, the absorbed light energy is distributed exponentially along the thickness direction, indicating the surface specificity of the evanescent wave probe. If we define the depth as the distance at which the evanescent field intensity decays by a factor of $1/e$, eqs 3.16 and 3.28 will be reduced to the commonly used depth-of-penetration equation for both s- and p-polarizations

$$d_p = \frac{\lambda}{2\pi\mathrm{Im}(\xi_2)} \qquad (3.33)$$

In the special case in which the sample is transparent, the penetration depth can be explicitly expressed as

$$d_p = \frac{\lambda}{2\pi[n_1^2\sin^2\alpha_1 - n_2^2]^{1/2}} \qquad (3.34)$$

Because d_p is a function of the wavelength, it is important to remember that in quantitative analysis taking the ratio of bands separated by a certain amount of energy may lead to significant misinterpretations. The relationships between the penetration depth and the wavelength of the light for KRS-5 and Ge are shown in Figure 3.4. Incidentally, using bands that were not corrected for optical effects and are close to each other usually leads to even more problems, related to the fact that there is significant dispersion around absorption bands. If two bands are close together, their refractive index components will be close enough to overlap, thus affecting the actual band intensities in such a way that the higher energy band will usually be less intense than the lower energy one. Figure 3.5 illustrates an example of two bands in which spectrum A is an ATR spectrum as recorded and spectrum B was subjected to corrections. Another example is Figure 3.6, which shows as-received (trace A) and corrected (trace B) portions of the ATR Fourier transform IR spectrum of 50/50 w/w% styrene/n-butyl acrylate. The band at 700 cm^{-1} is much more intense in trace A because light penetrates deeper and so the radiation encounters more sample.

Finally, the absorbed energy distributions expressed by eqs 3.31 and 3.32 can be related to the penetration depth as follows:

$$-\frac{\partial \langle s^{tr}_{z,\perp} \rangle}{\partial Z} = \frac{c \mathrm{Re}\xi_2 \mathrm{Im}\xi_2}{2\lambda} \left| U^{tr} \right|^2 e^{-\frac{z}{d_p/2}} \tag{3.35}$$

and

$$-\frac{\partial \langle s^{tr}_{z,\parallel} \rangle}{\partial Z} = \frac{c \mathrm{Re}\xi_2 / \hat{n}_2 \mathrm{Im}\xi_2}{2\lambda} \left| V^{tr} \right|^2 e^{-\frac{z}{d_p/2}} \tag{3.36}$$

Effective Thickness

As mentioned in Chapter 1, ATR spectroscopy was derived as a sampling technique for surface analysis, but the transmission measurements approach was almost exclusively used in spectral analysis. Therefore, before we focus on the details of sample thickness in ATR measurements, we begin with an analysis of the fundamental theory for a transmission experiment. At normal incidence along the z direction and ignoring reflectance, we can express the electromagnetic fields of light in a sample as

$$E = We^{\frac{i2\pi}{\lambda}z\hat{n}_2} e^{-i2\pi vt} \qquad 0 \le z \le d \tag{3.37}$$

and

$$H = \hat{n}_2 E \qquad 0 \le z \le d \tag{3.38}$$

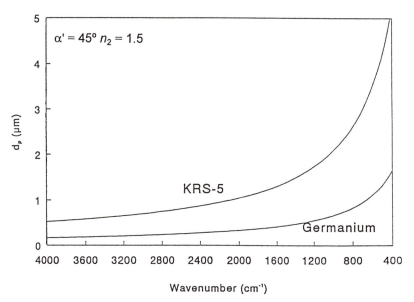

Figure 3.4. Relationship beween penetration depth and wavenumber for (A) KRS-5 crystal and (B) Ge crystal.

where d is the sample thickness. Using these expressions, we can calculate the energy flux in the sample as

$$\langle s_z \rangle = \frac{cv}{4\pi} \int_0^{1/v} \mathrm{Re}(E_y)\mathrm{Re}(H_x)\,dt$$
$$= \frac{c\mathrm{Re}(n_2)}{8\pi} W^2 e^{-\frac{2\pi}{\lambda}z\mathrm{Im}(n_2)} \tag{3.39}$$

Therefore, the transmittance will be

$$T = \frac{\langle s_z(d) \rangle}{\langle s_z(0) \rangle} = e^{-\beta_2 d} \tag{3.40}$$

where β_2 represents the absorbance per unit path length and is given by

$$\beta_2 = \frac{4\pi}{\lambda} k_2 \tag{3.41}$$

Equation 3.40 is the well-known Beer–Lambert law. Because of this relationship, the measured spectra are often expressed as absorbance (the negative logarithm of the transmittance), which is proportional to the absorption index k_2 and the path length d.

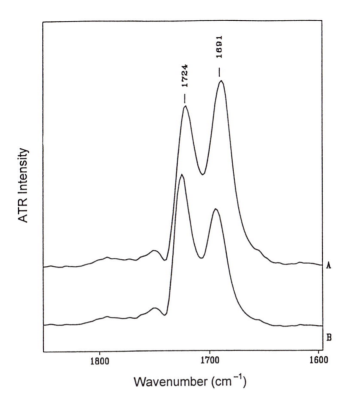

Figure 3.5. (A) ATR spectrum as received of polyol/isocyanate cross-linked polymer; (B) ATR spectrum corrected for optical effects.

Unlike the transmittance relationships used in transmission measurements, the reflectance expressions for a reflection experiment described by eqs 3.17 and 3.29 are not simple exponential functions of the absorption index. But even though they are complex functions of n_1, k_1, n_2, and k_2 and taking a negative logarithm of the reflectance has no physical significance, we can formulate the following analogous expression:

$$R = e^{-\beta_2 d_e} \qquad (3.42)$$

where d_e is the effective thickness—the film thickness of the sample that would give the same transmittance at normal incidence as that obtained in an ATR experiment. Apparently, the effective thickness, like R itself, is a complicated function, and the plot of $-\ln(R/\beta)$ as a function of β depends not only on optical conditions but also on the optical properties of the system. Figure 3.7 shows how $-\ln(R/\beta)$ values change as a function of β for a

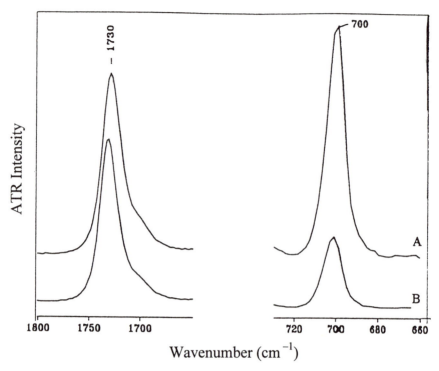

Figure 3.6. Two spectral regions of a 50/50 w/w% Styrene/n-butyl acrylate copolymer latex: (A) as received and (B) corrected for optical effects.

KRS-5 crystal. The first obvious conclusion is that for weak absorbances the $-\ln(R/\beta)$ values remain constant. However, as the absorbance approaches 1, the situation changes drastically. For Curve A, the $-\ln(R/\beta)$ values were calculated using the effective thickness equation for a Lorentzian band at 1700 cm^{-1} at an angle of incidence of 45°. Using the same parameters and a half-width-at-half-maximum of 3 cm^{-1}, Curve B represents maximum values obtained for the same band calculated using Fresnel's equation. Finally, Curve C was calculated using Fresnel's equation, but ignoring the dispersive behavior of the refractive index. Interestingly enough, when the KRS-5 crystal is replaced in these simulations by Ge, which has a higher refractive index, using the same spectral parameters and simulation procedures, the curves shown in Figure 3.8 were obtained. Regardless of shape of the plots for KRS-5 and Ge, one conclusion that can be drawn from these simulations is that with Ge, which provides shallower depths of penetration for a given angle of incidence, there is a greater tolerance for more intense bands. This is because the bands will be weaker, and for such

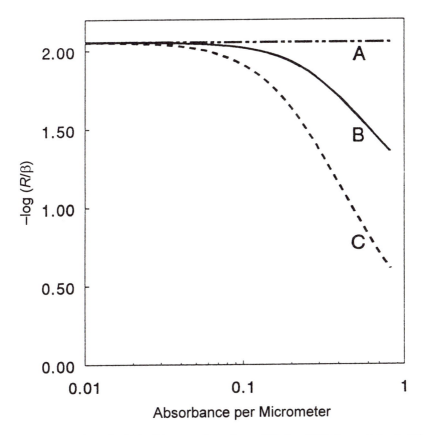

Figure 3.7. Plot of $-\ln(R/\beta)$ as a function of β for KRS-5 crystal at $45°$ angle of incidence: (A) calculated using the effective index equation for a band at 1700 cm^{-1}; (B) maximum values of the ATR intensity at 1500 cm^{-1} with half width at half maximum of 3 cm^{-1} calculated using Fresnel's equation; (C) same as B, but ignoring dispersion of the refractive index and using band at 1500 cm^{-1}. Other parameters: $n_0 = 2.38$, $\alpha = 45°$, and $n_\infty = 1.5$.

bands under total reflection conditions, where R is close to unity, the effective thickness may be approximated by

$$d_e = \frac{-\ln R}{\beta_2} \approx \frac{1-R}{\beta_2} \qquad (3.43)$$

The effective thickness for s-polarization can be obtained by inserting eq 3.17 into equation 3.43, giving

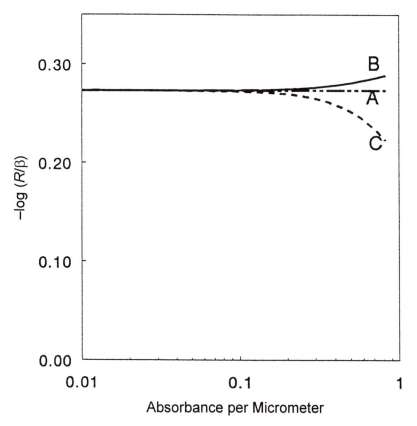

Figure 3.8. Plot of −ln(R/ β) as a function of β for Ge crystal at 45° angle of incidence: (A) calculated using the effective index equation for a band at 1700 cm⁻¹; (B) maximum values of the ATR intensity at 1700 cm⁻¹ with half width at half maximum of 3 cm⁻¹ calculated using Fresnel's equation; (C) same as B, but ignoring dispersion of the refractive index and using band at 1700 cm⁻¹. Other parameters: n_0 = 4.0, α = 45°, and n_∞ = 1.5.

$$d_{e,\perp} = \frac{1 - |\hat{r}|^2}{\beta_2}$$

$$= \frac{|\xi_1 + \xi_2|^2 - |\xi_1 - \xi_2|^2}{\beta_2 |\xi_1 + \xi_2|^2}$$

$$= \frac{2(\xi_1^* \xi_2 + \xi_1 \xi_2^*)}{\beta_2 |\xi_1 + \xi_2|^2} \tag{3.44}$$

$$= \frac{4\xi_1 \text{Re}(\xi_2)}{\beta_2 |\xi_1 + \xi_2|^2}$$

$$= \frac{4\xi_1 n_2 k_2}{\beta_2 \text{Im}(\xi_2) |\xi_1 + \xi_2|^2}$$

In these calculations, a mathematical identity between $k_2 n_2$ and $\text{Im}(\xi_2)\text{Re}(\xi_2)$ was used, and the asterisk indicates the conjugate of a complex number. If the intensities of the absorption bands are so low that $k_2 \ll n_2$ and $n_2 k_2 \ll n_2^2 - k_2^2 - n_1^2 \sin^2\alpha_1$, we can make the approximations

$$\xi_2 = \sqrt{n_2^2 - k_2^2 - n_1^2 \sin^2\alpha_1 + ik_2 n_2}$$

$$= \sqrt{n_2^2 - n_1^2 \sin^2\alpha_1} \tag{3.45}$$

$$= i\text{Im}(\xi_2)$$

and

$$|\xi_1 + \xi_2|^2 \approx \xi_1^2 + \text{Im}(\xi_2)^2 = n_1^2 - n_2^2 \tag{3.46}$$

We can substitute these expressions into eq 3.44 to obtain

$$d_{e,\perp} \approx \frac{\lambda n_2 n_1 \cos\alpha_1}{\pi(n_1^2 - n_2^2)\sqrt{n_1^2 \sin^2\alpha_1 - n_2^2}} \tag{3.47}$$

Similarly, we can obtain the expressions for the d_e values for p-polarized light by substituting eq 3.29 into eq 3.43:

Similarly, we can obtain the expressions for the d_e values for p-polarized light by substituting eq 3.29 into eq 3.43:

$$
\begin{aligned}
d_{e,\parallel} &= \frac{1 - |\hat{r}_\parallel|^2}{\beta_2} \\[2mm]
&= \frac{|\xi_1/n_1^2 + \xi_2/n_2^2|^2 - |\xi_1/n_1^2 - \xi_2/n_2^2|^2}{\beta_2|\xi_1/n_1^2 + \xi_2/n_2^2|^2} \\[2mm]
&= \frac{2((\xi_1/n_1^2)^*(\xi_2/n_2^2) + (\xi_1/n_1^2)(\xi_2/n_2^2)^*)}{\beta_2|\xi_1/n_1^2 + \xi_2/n_2^2|^2} \\[2mm]
&= \frac{4(\xi_1/n_1^2)\mathrm{Re}(\xi_2/n_2^2)}{\beta_2|\xi_1/n_1^2 + \xi_2/n_2^2|^2}
\end{aligned}
\tag{3.48}
$$

Using weak band approximations, we obtain

$$
\begin{aligned}
\mathrm{Re}(\xi_2/n_2^2) &= \frac{(n_2^2 - k_2^2)\mathrm{Re}(\xi_2) + 2n_2 k_2 \mathrm{Im}(\xi_2)}{(n_2^2 - k_2^2)^2 + 4n_2^2 k_2^2} \\[2mm]
&\approx \frac{n_2^2 \mathrm{Re}(\xi_2)\mathrm{Im}(\xi_2) + 2n_2 k_2[\mathrm{Im}(\xi_2)]^2}{\mathrm{Im}(\xi_2)n_2^4} \\[2mm]
&\approx \frac{n_2 k_2(2n_1^2 \sin^2\alpha_1 - n_2^2)}{n_2^4(n_1^2 \sin^2\alpha_1 - n_2^2)^{1/2}}
\end{aligned}
\tag{3.49}
$$

and

$$
\begin{aligned}
\left|\frac{\xi_1}{n_1^2} + \frac{\xi_2}{n_2^2}\right|^2 &\approx \left(\frac{\xi_1}{n_1^2}\right)^2 + \left[\frac{\mathrm{Im}(\xi_2)}{n_2^2}\right]^2 \\[2mm]
&= \frac{1}{n_1^2 n_2^2}[n_2^4 \cos^2\alpha_1 + n_1^2(n_1^2 \sin^2\alpha_1 - n_2^2)] \\[2mm]
&= \frac{1}{n_1^2 n_2^2}[n_2^4 - n_1^2 n_2^2 + (n_1^4 - n_2^4)\sin^2\alpha_1] \\[2mm]
&= \frac{1}{n_1^2 n_2^2}(n_1^2 - n_2^2)[(n_1^2 + n_2^2)\sin^2\alpha_1 - n_2^2]
\end{aligned}
\tag{3.50}
$$

and combining eqs 3.48 through 3.50 gives the following relationship:

$$d_{e,\parallel} = \frac{\lambda(\cos\alpha_1)n_1 n_2[2n_1^2\sin^2\alpha_1 - n_2^2]}{\pi[n_1^2\sin^2\alpha_1 - n_2^2]^{1/2}(n_1^2 - n_2^2)[(n_1^2 + n_2^2)\sin^2\alpha_1 - n_2^2]} \tag{3.51}$$

Figures 3.9 and 3.10 illustrate how the film effective thickness changes as a function of polarization and with the choice of the ATR crystal. The p-polarized light always gives a larger effective thickness for a given angle of incidence, and this difference becomes more pronounced as the critical angle is approached. The curves in Figure 3.9 end around the incidence angle of 40° because, as shown in Figure 1.5, the critical angle for KRS-5 for a sample with a refractive index of $n_2 = 1.5$ is about 40°.

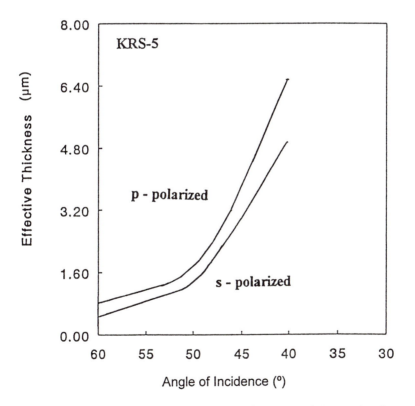

Figure 3.9. Effective thickness plotted as a function of the angle of incidence for KRS-5 crystal for p- and s-polarized light. Both curves were generated with the following parameters: $n_1 = 2.38$ and $n_2 = 1.5$.

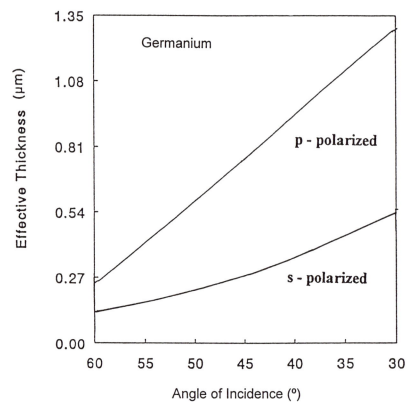

Figure 3.10. Effective thickness plotted as a function of the angle of inci-dence for Ge crystal for p- and s-polarized light. Both curves were generated with the following parameters: n_1 = 2.38 and n_2 = 1.5.

References

1. Harrick, N. J. *J. Phys. Chem.* **1960**, *64*, 1110.

2. Fahrenfort, J. *Spectrochim. Acta* **1961**, *17*, 698.

3. Sharpe, L. H. *Proc. Chem. Soc.* **1961**, Dec., 461.

4. Hirayama, T.; Urban, M. W. *Prog. Org. Coat.* **1992**, *21*, 81.

5. Tiefenthaler, A. M.; Urban, M. W. *Appl. Spectrosc.* **1988**, *42*, 163.

6. Evanson, K. W.; Urban, M. W. *J. Appl. Polym. Sci.* **1991**, *42*, 7.

7. Evanson, K. W.; Thorstenson, T. A.; Urban, M. W. *J. Appl. Polym. Sci.* **1991**, *42*, 2297.

8. Evanson, K. W.; Urban, M. W. *J. Appl. Polym. Sci.* **1991**, *42*, 2309.

9. Mirabella, F. M. *Appl. Spectrosc. Rev.* **1985**, *21*, 45.

10. Urban, M. W.; Koenig, J. L. In *Applications of FT-IR Spectroscopy*; Durig, J. R., Ed.; Vibrational Spectra and Structure, Vol. 18; Elsevier: Amsterdam, 1990; Chapter 3.

11. Tshmel, A. E.; Vettegren, V. I.; Zolotarev, V. M.; Ioffe, A. F. *Macromol. Sci. Phys.* **1982**, B21, 243.

12. Harrick, N. J. *Internal Reflection Spectroscopy*; Interscience: New York, 1967.

13. Sung, C. S. P. *Macromolecules* **1981**, *14*, 591.

14. Hobbs, J. P.; Sung, C. S. P.; Krishnan, K.; Hill, S. *Macromolecules* **1983**, *16*, 193.

4

Transmission Versus ATR Spectra

Differences Between Transmission and ATR Spectra

In the previous chapters, we saw that the optical dispersion effect may have a strong influence on the band intensities in ATR spectra. In this chapter, we will use several simulated transmission and ATR spectra to examine how optical effects can influence spectral resolution and intensity changes. As was pointed out earlier, the differences between ATR and transmission spectra arise from the experimental conditions, as discussed by Harrick (*1*) and Graf et al. (*2*). These studies clearly concluded that the frequency shifts—along with the sensitivity difference between high- and low-wavenumber regions (*1*) and the spectral distortion at angles near the critical angle—are the primary concerns in ATR quantitative analysis. Although most of these concerns can be addressed, the effect of optical dispersion is more complex and can be only partially avoided as long as the angle of incidence is a few degrees higher than the critical angle and only weak bands are a subject of analysis (*1*). For overlapping bands, however, the optical dispersion effect can cause considerable changes in the relative intensities of the overlapping bands, and ignoring these effects may lead to improper interpretation of the ATR results. Perhaps the most accurate statement made about ATR was by Jack Koenig:

> Because ATR is a contact method, it is generally believed that the technique is a surface analysis technique, and if you believe that a penetration of 1 μm is probing the surface, then ATR is such a method. However, because frequency shifts are observed as a result of the optical physics of the system, it has often been suggested that the IR spectra of some samples have frequencies that are different for the surface species than for the bulk sample. Well,

3348–9/96/0049/$15.25/0/© 1996 American Chemical Society

I certainly believe that this is possible, but generally the observations are a result of IRS [internal reflection spectroscopy]-induced frequency shifts rather than differences of a chemical nature. (*3*)

In an effort to address the issues of optical effects, attempts have been made to resolve the ATR spectral distortion by using asymmetric curve fitting (*4*), but these approaches are essentially mathematical manipulations and do not take into account the optical dispersion effects that actually cause the spectral distortion. As was identified in the literature, one approach to eliminating the optical dispersion effects is to convert an ATR spectrum into absorbance units using the Kramers–Kronig relationship (*5–7*). While the Kramers–Kronig relationship will be discussed extensively in the next chapter, here we will discuss the implications of not using it.

According to the Beer-Lambert law, the band intensity is proportional to the concentration of a given species, and the proportionality constant is the band absorption coefficient. The absorption coefficient is wavenumber-dependent, and the accuracy of its determination may be affected by the method used. However, quantitative analysis of the IR bands can be performed only if the absorption coefficient of the band of interest is established. This approach requires measuring the IR band intensities as a function of concentration. In practice, however, the band of interest may often overlap with other bands caused by the same or other components in a mixture. As a matter of fact, for a spectroscopist, the presence of two or more components is not an uncommon situation and requires ratioing of band intensities to establish the relative concentrations in the mixture (*8, 9*). Although these assessments can be safely used when IR spectra are recorded in transmission mode, the situation changes drastically when surfaces are analyzed by optical methods. In many applications of ATR Fourier transform IR spectroscopy (*10–12*), the approaches used for transmission measurements have been applied inappropriately. Although the issue of overlapping bands in ATR spectra is not a new one and has been addressed by Hawranek and Jones (*13*), the primary focus was on the errors caused by truncation of overlapping bands while using numerical Kramers–Kronig transformations. In the following sections, we will examine the validity of these assumptions, especially when the bands of interest overlap. In this context, it is important to establish how band intensities and separation between bands of interest affect ATR analysis.

Generating Transmission and ATR Spectra

Because experimental conditions for transmission and ATR experiments are so different, it is a common practice to simulate transmission and ATR spectra and, upon establishing a testing method, examine the effect of band separation and band intensities on both spectra. Following the Beer–Lambert law (eq 1.2), the absorbance spectrum is expressed in absorbance units per micrometer, whereas the ATR single-reflection spectrum is presented as the negative logarithm of reflectance. For the absorbance spectrum, $\beta_2(\tilde{v})$, the band intensities depend on the absorption and refractive indices of the sample, $k_2(\tilde{v})$ and $n_2(\tilde{v})$, and can be expressed by

$$\beta_2(\tilde{v}) = 4\pi\,\tilde{v}k_2(\tilde{v}) \times 10^{-4} \tag{4.1}$$

and

$$k_2(\tilde{v}) = \frac{-2\tilde{v}}{\pi}\int_0^{+\infty} \frac{n_2(s)}{s^2 - \tilde{v}^2}\, ds \tag{4.2}$$

where $\tilde{v}$ is wavenumber (cm^{-1}) and s is the integral variable (14). Its ATR counterpart for s-polarized incident light (TE), $\hat{r}(\tilde{v})$, is given by Fresnel's equation (1).

$$\hat{r}(\tilde{v}) = -\ln\left|\frac{n_1\cos\alpha_1 - [(n_2^2 - ik_2^2)^2 - n_1^2\sin^2\alpha_1]^{1/2}}{n_1\cos\alpha_1 + [(n_2^2 - ik_2^2)^2 - n_1^2\sin^2\alpha_1]^{1/2}}\right|^2 \tag{4.3}$$

where n_1 is the refractive index of the ATR crystal and α_1 is the angle of incidence. Because the refractive index spectrum exhibits substantial dispersion around the absorption bands, its magnitude depends strongly on wavenumber (1). If the refractive index at the infinite wavenumber ($n_{2,\infty}$) is known, the wavenumber-dependent refractive index can be calculated from eq 4.2, using the Kramers–Kronig relationship (15).

To illustrate how overlapping bands affect the intensities of the bands recorded in transmission (absorbance) and ATR modes of detection, two absorption bands, designated as a and b, will be generated using the following function:

$$f_i(\tilde{v}) = \frac{S_i\omega_i\tilde{v}/\tilde{v}_{m,i}}{[1 - (\tilde{v}/\tilde{v}_{m,i})^2]^2 + \omega_i^2(\tilde{v}/\tilde{v}_{m,i})^2} \tag{4.4}$$

where $i = a$ or b. Although other functions, such as Gaussian or Lorentzian functions, can be used as well, this choice was dictated by the fact that the bands generated by this function exhibit a maximum at $\tilde{v} = \tilde{v}_m$ equal to S/ω and the half-width at half-maximum (HWHM) divided by $\tilde{v}_m$ is a monotonic function of ω. For $\omega \ll 1$, HWHM $\approx \omega\tilde{v}_m/2$. Furthermore, the use of eq 4.4 is beneficial because the corresponding refractive index spectrum can be calculated without using numerical integration. Therefore, the exact analytical solution can be obtained by taking the Kramers–Kronig transform of $f_i(\tilde{v})$ in eq 4.4 (16) to obtain

$$g_i(\tilde{v}) = \frac{S_i[1 - (\tilde{v}/\tilde{v}_{m,i})^2]}{[1 - (\tilde{v}/\tilde{v}_{m,i})^2]^2 + \omega_i^2(\tilde{v}/\tilde{v}_{m,i})^2} \qquad (4.5)$$

Adding two bands given by eq 4.4 and generated at different values of $\tilde{v}_m$ gives the absorption index spectrum $k_2(\tilde{v})$, whereas the refractive index spectrum can be obtained by adding the Kramers–Kronig transforms of eq 4.4 generated for bands a and b to the refractive index at the infinite wavenumber $n_{2,\infty}$. Both k_2 and n_2 spectra are illustrated in Figure 4.1. For simplicity, S, ω, and $\tilde{v}_m$ in eq 4.4 are chosen so that the two bands in the absorbance spectrum are located symmetrically on both sides of a fixed wavenumber $\tilde{v}_0$ and have the same intensity (β_{max}) and the same HWHM. This is accomplished by letting $\omega_a = \omega_b = \omega$, $\tilde{v}_{m,a} = \tilde{v}_0 + \Delta\tilde{v}/2$; $\tilde{v}_{m,b} = \tilde{v}_0 - \Delta\tilde{v}/2$, $S_a = k_{max}\omega\tilde{v}_0/\tilde{v}_{m,a}$, and $S_b = k_{max}\omega\tilde{v}_0/\tilde{v}_{m,b}$, where $\Delta\tilde{v}$ is the separation between a and b bands and $k_{max} = \beta_{max}/4\pi\tilde{v}_0$. For convenience, all spectra were simulated using $\tilde{v}_0 = 1500$ cm^{-1}, $\omega = 0.002$ (HWHM ≈ 1.5 cm^{-1}), $\alpha_1 = 45°$, $n_1 = 2.38$, and $n_{2,\infty} = 1.5$. Once the spectral parameters are defined, we can focus on the effect of band separation for s- and p-polarized spectra.

Effect of Band Separation in s-Polarized Spectra

To examine how band separation may affect the relative intensities of simulated bands a and b, the procedures described in the previous section should be employed. A series of spectra generated by adding a and b bands separated by different values of $\Delta\tilde{v}$ is shown in Figure 4.2. Although the bands separated by more than 4 cm^{-1} can be resolved, trace A appears as one band because the separation of 2 cm^{-1} is smaller than twice the HWHM. Traces B, C, D, and E illustrate the same bands ($k_{max} = 0.25$) separated by 4, 6, 10, and 18 cm^{-1}, respectively. As seen, the relative band intensities do not change with separation. In contrast, Figure 4.3 shows the same simulated spectra for ATR bands. In this case, the intensity maxima of the two bands are no longer the same, and the band shapes are considerably distorted. This is further illustrated in Figure 4.4, where the ATR spectra A_1, B_1, and C_1 are overlaid with the corresponding transmission spectra A_2, B_2, and C_2. From these results, it appears that at smaller band separations, the band HWHM limit does not allow us to resolve a and b bands. Furthermore, a frequency shift is observed. When the band separation between a and b in the absorbance spectrum (Figure 4.4, trace B_2) exceeds twice the HWHM, a double band appears. This is not the case for the ATR spectrum in trace B_1, where band a appears as a shoulder on the more intense band b. For $\Delta\tilde{v} = 6$ cm^{-1} (traces C_1 and C_2), the bands are fairly

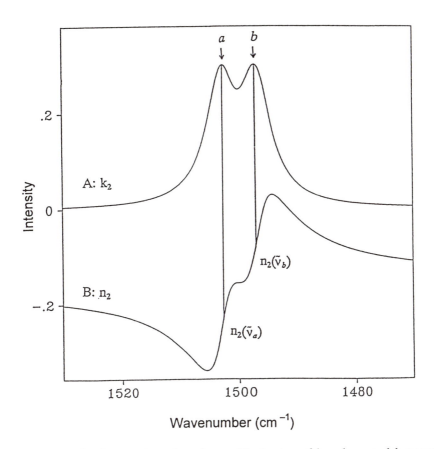

Figure 4.1. k_2 (trace A) and n_2 (trace B) spectra of bands a and b near 1500 cm⁻¹.

well resolved in both ATR and absorbance spectra, but the relative band intensities in the ATR spectra are still considerably different.

In an effort to establish the effect of band separation on intensity, a plot of the maximum intensity of *a* and *b* in each ATR spectrum as a function of the band separation was constructed. As shown in Figure 4.5, with increasing separation between *a* and *b*, the intensity of band *a* decreases sharply to minimum and then gradually increases. In contrast, the intensity of band *b* decreases monotonically. When the band separation exceeds approximately 30 cm⁻¹, or 20 times the HWHM, the intensity changes of both bands level off, indicating that for such defined spectral conditions, only bands that are separated by 30 cm⁻¹ will not affect one another. Therefore, spectral distortions are anticipated that, as we will see later, can be eliminated by using proper algorithms.

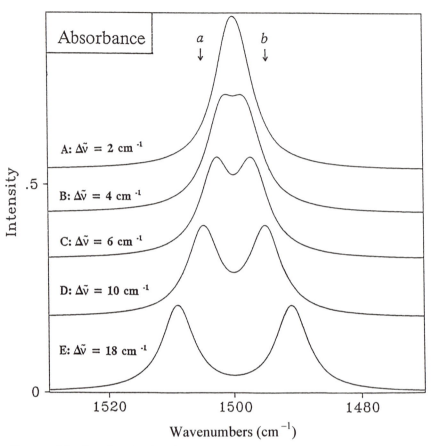

Figure 4.2. Simulated absorbance spectra with k_{max} = 0.25. (A) Δv = 2 cm^{-1}, (B) $\Delta \tilde{v}$ = 4 cm^{-1}, (C) $\Delta \tilde{v}$ = 6 cm^{-1}, (D) $\Delta \tilde{v}$ = 10 cm^{-1}, and (E) $\Delta \tilde{v}$ = 18 cm^{-1}.

Effect of Band Intensities for s-Polarized Spectra

Let us now determine how the intensity ratio of bands *a* and *b* is affected by their separation. This relationship is plotted in Figure 4.6, and curves A and B represent the intensity ratios for absorbance (transmission) and ATR spectral conditions, respectively. In the ATR spectra, the intensity ratio decreases sharply to a minimum, and then gradually increases. At the same time, the absorbance band ratio remains constant, and the differences between absorbance and ATR band ratios are more pronounced at smaller band separations. Note also that the band separation below 60 cm^{-1} affects the band intensity ratios. For larger band separations, however, the factor of $1/\lambda$ in the expression for penetration depth (*11*) in ATR experiments

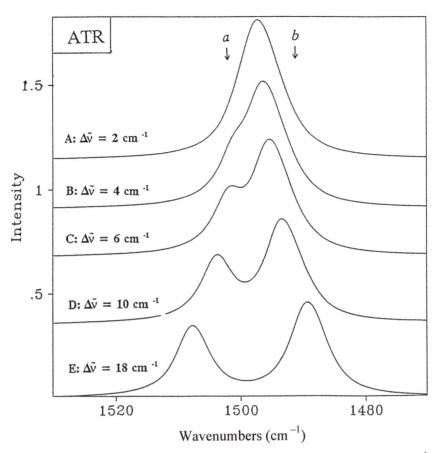

Figure 4.3. Simulated ATR spectra with k_{max} = 0.25. (A) Δv = 2 cm^{-1}, *(B)* $\Delta \tilde{v}$ = 4 cm^{-1}, *(C)* $\Delta \tilde{v}$ = 6 cm^{-1}, *(D)* $\Delta \tilde{v}$ = 10 cm^{-1}, *and (E)* $\Delta \tilde{v}$ = 18 cm^{-1}.

may become dominant. Therefore, the difference between the ATR and absorbance intensities will increase even further at greater band separations. This simulation may have various consequences for quantitative analysis using ATR measurements. For example, if a band chosen for normalization is too close to the band of interest, perturbation between adjacent bands may occur, making quantitative analysis troublesome. In contrast, if the band used for normalization is far from the band of interest, we need to take into account the dependence of the penetration depth on wavenumber (Figure 3.4). Under these circumstances, a rule of thumb is to choose a normalization band at some distance from the band of interest, about 20 to 30 times the band HWHM. Because such situations do not occur often, a commonly accepted approach is to convert the ATR

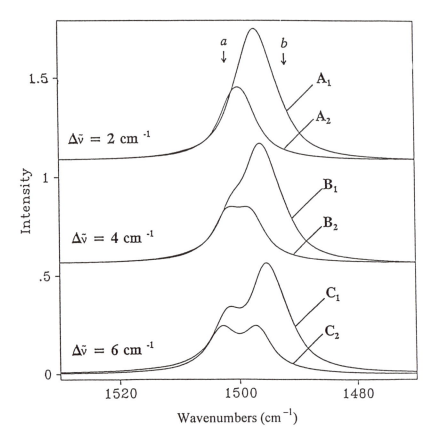

Figure 4.4. Simulated ATR spectra (A_1, B_1, and C_1) overlaid with the corresponding absorbance spectra (A_2, B_2, and C_2) for $\Delta\tilde{v}$ = 2, 4, and 6 cm^{-1}.

spectrum into the absorbance spectrum (4, 5). Such conversions, which are described in detail in Chapter 6, account for optical dispersion effects and the wavenumber dependence of the penetration depth, allowing the use of the Beer-Lambert law for quantitative analysis.

Let us now examine how the band intensity may affect the relative intensity changes of overlapping bands *a* and *b*. Figure 4.7 shows simulated absorbance spectra: each spectrum contains two absorption bands *a* and *b*, separated by $\Delta\tilde{v}$ = 6 cm^{-1}. The absorption intensity ratios are not affected. Figure 4.8 shows the simulated ATR spectra using the same optical constants, and as in previous ATR spectra, the relative band intensities change as the band intensity increases. Postponing for now discussion of the origin of these differences, let us estimate how the band intensity and band separation will affect ATR s- and p-polarized spectra.

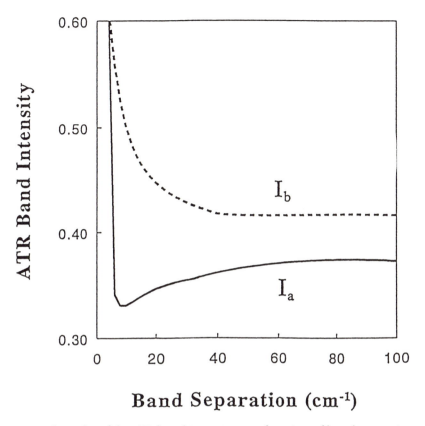

Figure 4.5. Plot of the ATR band intensity as a function of band separation.

Combined Effects of Band Intensity and Band Separation for s-Polarized Spectra

In an effort to establish how band separation may affect ATR band intensity, Figure 4.9 was constructed to show the relationship between the ATR band intensities and maximum absorbance. Curves A, B, and C represent the ATR intensities of band *b* obtained for separations of 6, 10, and 18 cm^{-1}, respectively, whereas curves A′, B′, and C′ are the corresponding plots for band *a*. These plots illustrate that for a given band separation, the intensity difference between bands *a* and *b* increases with increasing band intensity, and the increase is more pronounced as the band separation decreases. Figure 4.10 shows the intensity ratio of *a* and *b* (I_a/I_b), plotted as a function of the absorbance intensity. Curves A, B, and C were obtained for bands *a* and *b* separated by 6, 10, and 18 cm^{-1}, respectively, whereas straight line D represents the intensity ratio obtained from

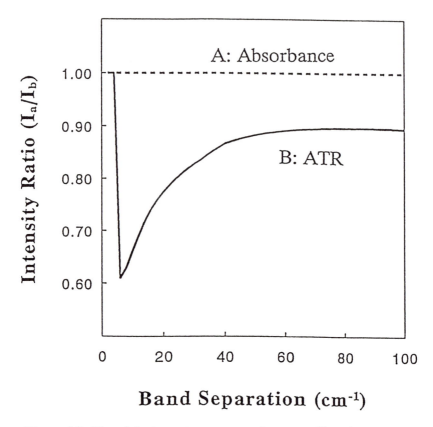

Figure 4.6. Plot of the intensity ratio as a function of band separation.

absorbance spectra (Figure 4.7). These results have numerous practical implications. Suppose that the bands *a* and *b* are absorption bands due to two components in an unknown mixture. According to the Beer–Lambert law, the horizontal line D in Figure 4.10 indicates an unchanging concentration ratio of the two components, whereas curves A, B, and C illustrate a nonlinear relationship between the ATR intensity ratio and the concentration ratio in a mixture.

Effect of Overlapping Bands on ATR Intensities for p-Polarization

Figures 4.11 and 4.12 are constructed for p-polarization. Although the effects of overlapping bands on ATR intensities are similar for s- and p-polarizations, the intensity effect for p-polarization is greater. For

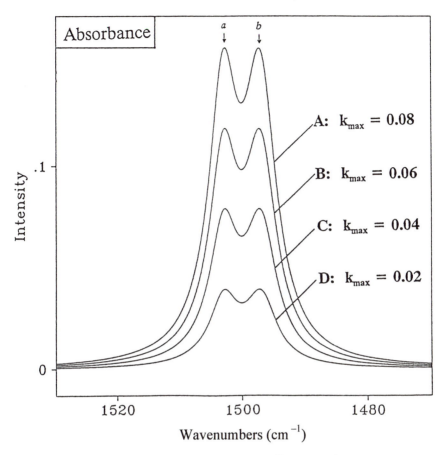

Figure 4.7. Simulated absorbance spectra for $\Delta\tilde{v}$ = 6 cm^{-1} and band intensities (A) k_{max} = 0.8, (B) k_{max} = 0.6, (C) k_{max} = 0.4, and (D) k_{max} = 0.2.

isotropic samples, the relative sensitivity of the p- and s-polarizations may be illustrated by examining the following forms of Fresnel's equation:

$$\hat{r}_\perp = \frac{\sin(\alpha_2 - \alpha_1)}{\sin(\alpha_2 + \alpha_1)} \tag{4.6}$$

and

$$\hat{r}_\parallel = \frac{\tan(\alpha_2 - \alpha_1)}{\tan(\alpha_2 + \alpha_1)} \tag{4.7}$$

where α_1 is the angle between the normal line of the interface and the direction of the light beam penetrating into the sample and α_2 is the angle of the transmitted beam (Figure 2.1). Combining eqs 4.6 and 4.7 yields

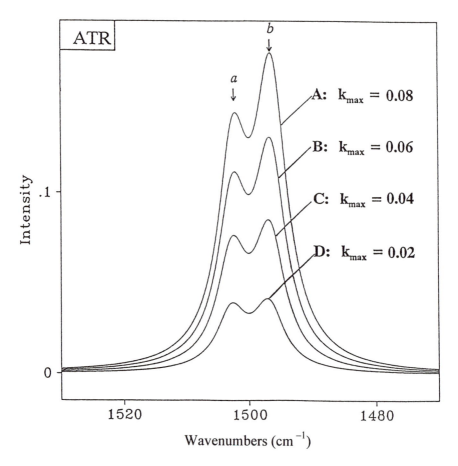

Figure 4.8. Simulated ATR spectra for $\Delta \tilde{v}$ = 6 cm^{-1} and band intensities (A) k_{max} = 0.8, (B) k_{max} = 0.6, (C) k_{max} = 0.4, and (D) k_{max} = 0.2.

$$\hat{r}_{\parallel} = \hat{r}_{\perp} \frac{\cos(\alpha_2 + \alpha_1)}{\cos(\alpha_2 - \alpha_1)}$$

$$= \hat{r}_{\perp} \frac{\sin(\pi/2 - \alpha_2 - \alpha_1)}{\sin(\pi/2 - \alpha_2 + \alpha_1)} \tag{4.8}$$

If we now assume an angle of incidence of 45° and use any of the crystals listed in Table 4.1 in the Appendix, eq 4.8 can be simplified into the following forms:

$$\hat{r}_{\parallel} = \hat{r}_{\perp}^2 \tag{4.9}$$

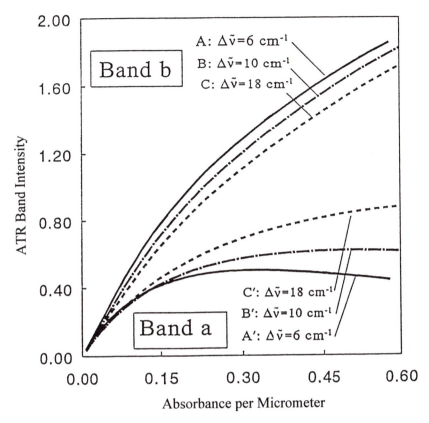

Figure 4.9. Plot of the ATR band intensity as a function of absorbance for the following band separations: (A and A') 6 cm^{-1}, (B and B') 10 cm^{-1}, (C and C') 18 cm^{-1}.

or using reflectance for p- and s-polarized light, $R_{\parallel}$ and $R_{\perp}$, respectively,

$$-\ln R_{\parallel} = -2 \ln R_{\perp} \qquad (4.10)$$

Equation 4.10 indicates that at 45°, the ATR intensity of the p-polarization is exactly twice the ATR intensity of the s-polarization. To illustrate the relative intensity changes in the p- and s-polarized ATR spectra for other angles using KRS-5 and Ge crystals, Figures 4.13 and 4.14 were constructed. These results indicate that, as long as the sample of interest is isotropic, the intensity ratio for p- and s-polarized ATR bands is barely affected by the absolute absorbance values.

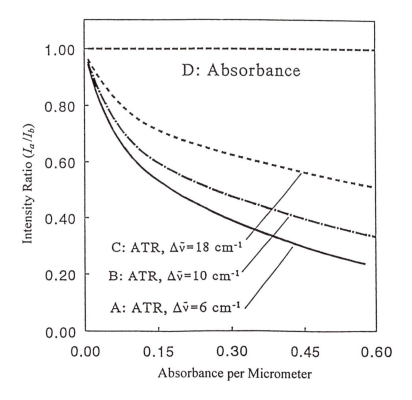

Figure 4.10. Plot of the intensity ratio as a function of absorbance for the following band separations: (A) 6 cm^{-1}, (B) 10 cm^{-1}, and (C) 18 cm^{-1}; (D) absorbance.

Origin of Differences and Practical Implications

At this point, it is appropriate to go back and once more address the issue of why the same absorbance bands located next to each other become so different in ATR spectra. According to Fresnel's equation (eq 4.3), the ATR band intensities increase with increasing $k_2(\tilde{v})$ and $n_2(\tilde{v})$ values. If $n_2(\tilde{v})$ values for bands a and b shown in Figure 4.1 were independent of wavenumber, two absorption bands with the same maximum $k_{2,\max}$ would yield two ATR bands with the same intensity. However, because of asymmetry of the optical dispersion (n_2) (trace B of Figure 4.1), the superposition of two dispersion bands results in a substantial difference between the refractive index values at the wavenumbers where the two absorption maxima occur.

To illustrate a practical implication of these well-known assessments, Figure 4.15 shows the measured ATR (trace A) and absorbance (trace B)

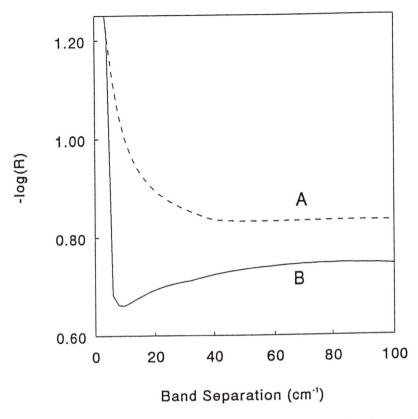

Figure 4.11. Plot of the p-polarized ATR band intensities of bands a and b as a function of band separation.

spectra of poly(ethyl acrylate) in the C–O stretching region. In this region, poly(ethyl acrylate) exhibits two absorption bands due to coupled asymmetric carbon–oxygen single-bond stretching vibrations. As in the simulated situations depicted by traces B_1 and B_2 of Figure 4.4, the C–O stretching band in the measured ATR spectrum (A) appears as a shoulder on a more intense band also due to C–O stretching. The measured absorbance spectrum appears to have two separated bands with similar intensities. The optical effects in the ATR spectrum can be corrected when the ATR spectrum is converted to the absorbance spectrum by using Kramers–Kronig transform procedures described in the literature (4, 5). Such procedures were once commonly used in the analysis of strong bands (17). The absorbance spectrum, obtained from the ATR spectrum (trace A) and shown in trace C, appears to have intensities similar to those in the absorbance spectrum (trace B) measured in a transmission experiment.

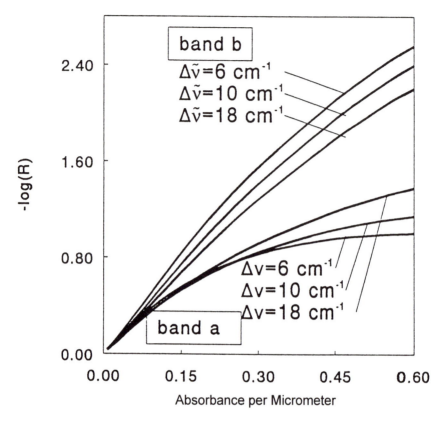

Figure 4.12. Plot of the p-polarized ATR band intensities as a function of absorbance for band separations of 6, 10, and 18 cm^{-1}.

These simulations show that in the analysis of overlapping bands there are significant differences between the band intensities in ATR and absorbance spectra. These differences are more pronounced for smaller band separations and stronger bands. The implications of optical effects are illustrated in Figures 4.16 and 4.17, which show two cases of "friendly" elastomer samples (friendly, because there is no problem in maintaining good contact between the sample and the crystal). Figure 4.16 depicts three spectra of freshly cut faces of polyurethane pellets. In the lab notebook of S. R. Skorich we read:

> In this experiment, six randomly selected pellets were sliced in half, exposing a pristine polymer surface representative of the bulk properties of the material. The same area coverage is achieved in all three cases, by placing the same number of pellet-halves at previously located evanescent wave maxima on the horizontal ATR crystals. The 60° Ge samples were taken first, then the 60° ZnSe,

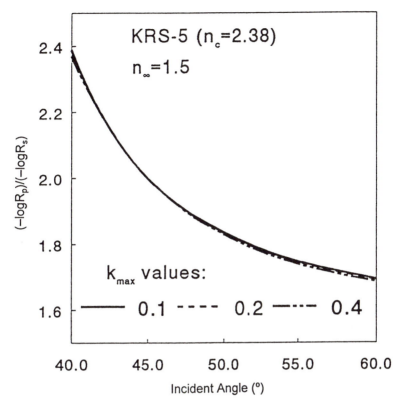

Figure 4.13. Intensity ratio of p- and s-polarized ATR spectra as a function of incident angle for KRS-5 crystal plotted for k_{max} values of 0.1, 0.2, and 0.4.

then the 45° ZnSe. This minimizes the possible spectral effect of chemical changes at the newly exposed surface. Since the probability of rapid and extensive chemical changes at the surface is considered to be low, these spectra are considered to show spectral changes due to predominantly the nature of the optical interface.

The wavenumber shifts for silicone elastomer spectra presented in Figure 4.17 are similar. These spectral changes with experimental conditions clearly tell us that the adoption of data-processing techniques including band ratioing and curve fitting may be well established for quantitative analysis of absorbance spectra, but their use for quantitative analysis of ATR spectra should be viewed with caution. ATR spectra should be converted to absorbance units by algorithms employing Kramers–Kronig transforms and Fresnel's relationships. This will be discussed in Chapter 6.

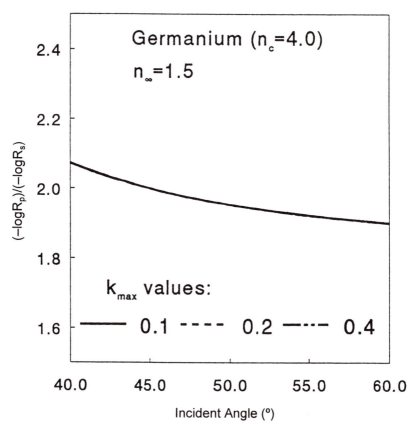

Figure 4.14. Intensity ratio of p- and s-polarized ATR spectra as a function of incident angle for Ge crystal plotted for k_{max} values of 0.1, 0.2, and 0.4.

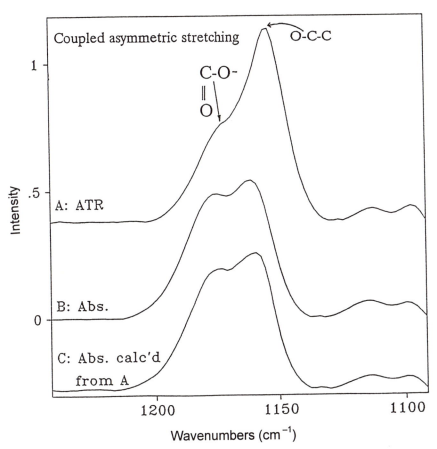

Figure 4.15. ATR and absorbance spectra of poly(ethyl acrylate): (A) experimental ATR Fourier transform IR spectrum; (B) experimental absorbance spectrum obtained in transmission measurements; (C) absorbance spectrum converted from ATR spectrum (trace A).

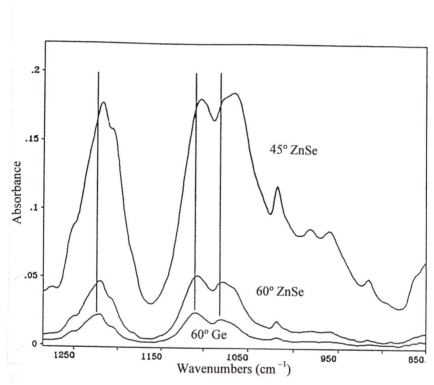

Figure 4.16. ATR Fourier transform IR spectra of freshly cut faces of poly-urethane elastomer pellets (courtesy S. R. Skorich, Medtronic Inc., private communication).

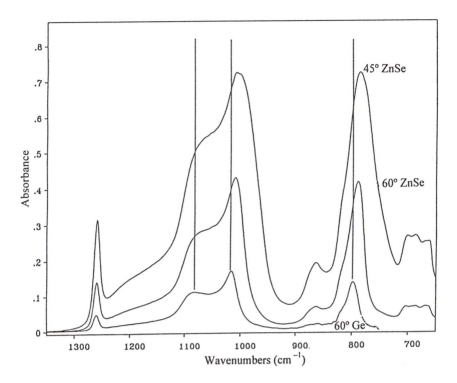

Figure 4.17. ATR Fourier transform IR spectra of silicone elastomer (courtesy S. R. Skorich, Medtronic Inc., private communication).

References

1. Harrick, N. J. *Internal Reflection Spectroscopy*; Interscience Publishers: New York, 1967.

2. Graf, R. T.; Koenig, J. L.; Ishida, J. In *Fourier Transmission Infrared Characterization of Polymers*; H. Ishida, Ed.; Plenum: New York, 1987.

3. Koenig, J. L. *Spectroscopy of Polymers*; American Chemical Society: Washington, DC, 1992.

4. Janik, L. J. *Appl. Spectrosc.* **1986**, *40*, 661.

5. Bertie, J. E.; Eysel, H. H. *Appl. Spectrosc.* **1985**, *39*, 392.

6. Bardwell, J. A.; Dignam, M. J. *Anal. Chim. Acta* **1986**, *181*, 253.

7. Goplen, T. G.; Cameron, D. G.; Jones, R. N. *Appl. Spectrosc.* **1980**, *34*, 652.

8. Hirshfield, T. B. *Anal. Chem.* **1976**, *48*, 721.

9. Koenig, J. L. *Adv. Poly. Sci.* **1983**, *54*, 87.

10. Popli, R.; Dwivedi, A. M. *J. Appl. Poly. Sci.* **1989**, *37*, 2469.

11. Zerbi, G.; Gallino, G.; Fanti, N. D.; Baini, L. *Polymer* **1989**, *30*, 2324.

12. Mirabella, J. F. M. *Spectroscopy* **1990**, *5*, 21.

13. Hawranek, J. P.; Jones, R. N. *Spectrochim. Acta* **1976**, *32A*, 99.

14. Fowles, G. R. *Introduction to Modern Optics*, 2nd ed.; Holt, Rinehart, and Winston: New York, 1975.

15. Crawford, B., Jr.; Goplen, T. G.; Swanson, D. in *Advances in Infrared and Raman Spectroscopy*; Clark, R. J. H.; Hester, R. E., Eds.; Heyden: London, 1980; Volume 4, Chapter 2.

16. Bardwell, J. A.; Dignam, M. J. *Anal. Chim. Acta* **1985**, *172*, 101.

17. Goplen, T. G.; Cameron, D. G.; Jones, R. N. *Appl. Spectrosc.* **1980**, *34*, 657.

Section II

Quantitative ATR Spectroscopy

5

Kramers–Kronig Transforms in ATR Spectroscopy

Introduction

Chapter 4 illustrated that the band separation and the band intensities drastically influence the shape and the relative ATR band intensities. Furthermore, as indicated in the first two chapters, in ATR experiments the measured reflectivity is a function of two unknown optical constants of the sample: the real (n_2) and imaginary (k_2) parts of a complex refractive index ($\hat{n}_2$). Although two or more measurements on the same sample under different experimental conditions may allow determination of both unknowns, theoretical considerations indicate that for a homogeneous surface a single measurement is enough because the two unknowns are related to each other. Such relations are called dispersion relations because they relate dispersive and absorptive processes. They are also referred to as Kramers–Kronig relations because H. A. Kramers (*1*) and R. de L. Kronig (*2, 3*) were the first to show dispersion relations for the dielectric constant and for the refractive index, with primary applications originally developed for dispersion of X-rays. This section presents the dispersion relations and their theoretical foundations for ATR analysis.

Kramers–Kronig Transforms

The general dispersion relations between the refractive and absorption index spectra are given by

$$n_2(\omega') = n_\infty + \frac{2}{\pi} P \int_0^{+\infty} \frac{\omega k(\omega)}{\omega^2 - \omega'^2} d\omega \tag{5.1}$$

and

3348–9/96/0073/$15.00/0/© 1996 American Chemical Society

$$k_2(\omega') = -\frac{2\omega}{\pi}P\int_0^{+\infty}\frac{n_2(\omega)}{\omega^2-\omega'^2}d\omega \tag{5.2}$$

where P is the principal value of the integral and ω is the angular frequency, which is related to wavenumber by $\omega = 2\pi\tilde{\nu}$. These two equations are mathematically equivalent to the following double Fourier transforms:

$$n_2(\omega) = n_\infty + 4\int_0^{+\infty}\cos(\omega t)\left[\int_0^{+\infty}k_2(\omega')\sin(\omega't)d\omega'\right]dt \tag{5.3}$$

and

$$k_2(\omega) = 4\int_0^{+\infty}\sin(\omega t)\left[\int_0^{+\infty}n_2(\omega')\cos(\omega't)d\omega'\right]dt \tag{5.4}$$

Although a combination of eqs 5.3 and 5.4 with Fresnel's equation may provide enough restrictive conditions to determine the optical constants, those constants cannot be expressed as an analytic function of the measured reflectivity. For this reason, theoretical descriptions in the literature (4–6) have focused on the relations between the amplitude and the phase of the Fresnel reflectivity. In the case of total internal reflection, the useful dispersion relation is

$$\theta(\omega') = \theta' + \frac{2\omega'}{\pi}P\int_0^{+\infty}\frac{\ln[R(\omega)^{1/2}]}{\omega^2-\omega'^2}d\omega \tag{5.5}$$

where θ' is given by

$$\theta' = 2\arctan\frac{[n_1^2\sin^2\alpha_1 - n_\infty^2]^{1/2}}{n_1\cos\alpha_1} \tag{5.6}$$

Given the reflectivity R from the ATR measurements and the phase angle θ calculated from eq 5.5, the complex-valued Fresnel reflection coefficient $\hat{r}$ can be obtained. This, in turn, allows direct calculation of the absorption and refractive indices from the following equations:

$$n_2 = n_1\mathrm{Re}\left[\sin^2\alpha_1 + \left(\frac{1-\hat{r}}{1+\hat{r}}\right)^2\cos^2\alpha_1\right]^{1/2} \tag{5.7}$$

and

$$k_2 = -n_1\mathrm{Im}\left[\sin^2\alpha_1 + \left(\frac{1-\hat{r}}{1+\hat{r}}\right)^2\cos^2\alpha_1\right]^{1/2} \tag{5.8}$$

Theory of Kramers–Kronig Transforms

All the dispersion relations are a result of the causality principle (7), which requires that no output precedes input. In spectroscopy that means that if there is no incident beam, no output in the form of a reflected beam can be obtained. Let $F(t)$ be a freely varying input starting at $t = 0$, with $G(t)$ being the corresponding output. However, for an infinitesimal input $F(t)dt$, in the period from t to $t + dt$, the output not only may exist during this period but may also extend to a later time t'. For example, the fact that the incident beam has been turned off does not necessarily mean that the reflecting beam is not received by a detector. Under such circumstances $dG(t)$ will be modified to, say, $dG'(t)$, and the contributions from $F(t)dt$ may be expressed as

$$dG(t') = K(t' - t)F(t)dt \qquad (5.9)$$

where $K(t' - t)$ is an output obtained from t to t'. The causality principle requires that $H(\tau) = 0$ at $\tau < 0$, where $H(\tau)$ is the linear response function. At this point, let us assume that the output signal is zero before the onset of the input and that after the input is turned off the output remains finite and decays to zero. With these assumptions, $H(\tau)$ is finite and approaches zero as τ approaches infinity. Adding all the infinitesimal contributions from 0 to t', and realizing that the contribution due to the input during the current period should be $H(0)F(t)dt/2$, we obtain

$$G(t') = G(0) + \int_0^{t'} [H(t' - t) - H(0)\delta(t' - t)/2]F(t)dt \qquad (5.10)$$

where $\delta(t' - t)$ is a Dirac function. Mathematically, eq 5.10 indicates that the output $G(t')$ is a convolution of the input $F(t)$ and the response kernel is $H(t) - H(0)\delta(t)/2$. Therefore, the Fourier transforms of these time variants are simply related by

$$g(\omega) = [h(\omega) - h_0]f(\omega) \qquad (5.11)$$

where $g(\omega)$, $h(\omega)$, and $f(\omega)$ are Fourier transforms of $G(t) - G(0)$, $H(t)$, and $F(t)$, respectively, and h_0 is a constant equal to $H(0)/2$. Of particular interest is the Fourier integral,

$$h(\omega) - h_0 = \int_{-\infty}^{+\infty} H(t)\exp(i\omega t)dt \qquad (5.12)$$

Because of the causality requirement that $H(\tau) = 0$ at $\tau < 0$, the lower limit of the above integral can be changed to 0, so

$$b(\omega) = b_0 + \int_0^{+\infty} H(t)\exp(i\omega t)dt \qquad (5.13)$$

This relationship along with the physically required reality of H(t) implies that the real and imaginary parts of $b(\omega)$ are given by

$$b_1 = b_0 + \int_0^{+\infty} H(t)\cos(\omega t)dt \qquad (5.14)$$

and

$$b_2 = \int_0^{+\infty} H(t)\sin(\omega t)dt \qquad (5.15)$$

It can be readily shown that $2ib_2(\omega)$ is the Fourier transform of an odd function $K(t) = H(t) - H(-t)$. For $t > 0$, $H(t) = K(t)$, and taking the inverse Fourier transform, we obtain

$$H(t) = K(t) = \int_{-\infty}^{+\infty} 2ib_2(\omega)\exp(-i\omega t)d\omega$$
$$\qquad\qquad\qquad\qquad t>0 \qquad (5.16)$$
$$= 4\int_0^{+\infty} b_2(\omega)\sin(\omega t)d\omega$$

Substituting eq 5.16 into eq 5.14 leads to

$$b_1(\omega) = b_0 + 4\int_0^{+\infty} \cos(\omega t)\left[\int_0^{+\infty} b_2(\omega')\sin(\omega' t)d\omega'\right]dt \qquad (5.17)$$

Alternatively, because $2b_1(\omega)$ is the Fourier transform of an even function $K'(t) = H(t) + H(-t)$, the following equation can be obtained:

$$b_2(\omega) = 4\int_0^{+\infty} \sin(\omega t)\left[\int_0^{+\infty} b_1(\omega')\cos(\omega' t)d\omega'\right]dt \qquad (5.18)$$

Equations 5.17 and 5.18 represent a dispersion relation expressed as a double Fourier transform.

We can use the causality requirement to derive a dispersion relation in a different form. If eq 5.13 is treated as the definition of $b(\omega)$ in the complex frequency domain $\omega = \omega_1 + i\omega_2$, $b(\omega)$ exhibits the following features:

1. The function is analytic in the upper half plane; that is, it contains no singularities there.

2. The limiting value of $b(\omega)$ is b_0 as ω_2 approaches infinity for any value of ω_1.

3. The real part of $b(\omega)$ is even, and the imaginary part is odd.

4. On the imaginary axis, where $\omega_1 = 0$, $b(\omega)$ is real and decreases monotonically from $b(0)$ to b_0 as ω_2 changes from 0 to infinity.

The analyticity of $b(\omega)$ guarantees a zero value of the following contour integral:

$$I = \oint \frac{b(\omega)}{\omega - \omega'}\, d\omega = 0 \qquad (5.19)$$

where ω' is a real constant and the contour is shown in Figure 5.1. By definition, the limiting value of the integral along the segments A and C as the radius of the large semicircle approaches infinity and that of the small semicircle approaches zero is the Cauchy principle value of the following integral:

$$I_{AC} = P\int_{-\infty}^{+\infty} \frac{b(\omega_1)}{\omega_1 - \omega'}\, d\omega_1 \qquad (5.20)$$

The integral along segment D may be written as

$$I_D = \int_0^\pi \frac{b(R_D e^{i\phi})}{R_D e^{i\phi} - \omega'}\, d(R_D e^{i\phi}) \qquad (5.21)$$

On segment D for $0 < \phi < \pi$, $\omega_2 = R_D \cos\phi$ approaches infinity as R_D approaches infinity. At this limit, the integrand of the above integral is b_0, and I_D becomes $i\pi b_0$. The integral along the semicircle B may be evaluated as follows:

$$
\begin{aligned}
I_D &= -\int_0^\pi \frac{b(\omega' + R_B e^{i\phi})}{R_B e^{i\phi}}\, d(\omega' + R_B e^{i\phi}) \\
&= -i\int_0^\pi b(\omega' + R_B e^{i\phi})\, d\phi
\end{aligned}
\qquad (5.22)
$$

As R_B approaches zero, the integrand approaches $b(\omega')$ and I_B approaches $-i\pi b(\omega')$. According to eq 5.17, $I = I_{AC} + I_B + I_D = 0$, and therefore we have

$$
\begin{aligned}
b(\omega') - b_0 &= b_1(\omega') + ib_2(\omega') - b_0 \\
&= P\int_{-\infty}^{+\infty} \frac{(-i/\pi)b_1(\omega_1) + (1/\pi)b_2(\omega_1)}{\omega_1 - \omega'}\, d\omega_1
\end{aligned}
\qquad (5.23)
$$

which can be separated into the following real equations:

$$b_1(\omega') - b_0 = \frac{1}{\pi}P\int_{-\infty}^{+\infty} \frac{b_2(\omega_1)}{\omega_1 - \omega'}\, d\omega_1 \qquad (5.24)$$

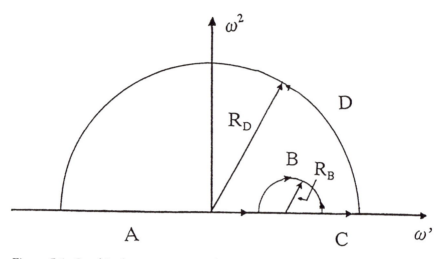

Figure 5.1. Graphical representation of separating the real and imaginary parts.

and

$$b_2(\omega') = -\frac{1}{\pi}P\int_{-\infty}^{+\infty}\frac{b_1(\omega_1)}{\omega_1-\omega'}\,d\omega_1 \qquad (5.25)$$

Because $b_1(\omega)$ is even and $b_2(\omega)$ is odd, eqs 5.24 and 5.25 can be transformed into integrals over positive frequency only:

$$b_1(\omega') = \frac{2}{\pi}P\int_0^{+\infty}\frac{\omega_1 b_2(\omega_1)}{\omega_1^2-\omega'^2}\,d\omega_1 \qquad (5.26)$$

and

$$b_2(\omega') = -\frac{2\omega'}{\pi}P\int_0^{+\infty}\frac{b_1(\omega_1)}{\omega_1^2-\omega'^2}\,d\omega_1 \qquad (5.27)$$

We can consider the electromagnetic field as an input; when the light passes through a polarizer, the polarized output is generated (8). In the frequency domain, the response function is simply the dielectric constant: $\varepsilon(\omega) - \varepsilon_0 = \varepsilon_1(\omega) + i\varepsilon_2(\omega) - \varepsilon_0$. Applying eqs 5.26 and 5.27, we obtain the dispersion relation between $\varepsilon_1(\omega)$ and $\varepsilon_2(\omega)$:

$$\varepsilon_1(\omega) = \varepsilon_0 + \frac{2}{\pi}P\int_{-\infty}^{+\infty}\frac{\omega_1\varepsilon_2(\omega_1)}{\omega_1^2-\omega^2}\,d\omega_1 \qquad (5.28)$$

and

$$\varepsilon_2(\omega) = -\frac{2\omega}{\pi} P \int_{-\infty}^{+\infty} \frac{\varepsilon_1(\omega_1)}{\omega_1^2 - \omega^2} \, d\omega_1 \qquad (5.29)$$

A comparison of eq 5.26 with eq 5.28, and eq 5.27 with eq 5.29, clearly indicates that the relation between the refractive index and the dielectric constant $\hat{n}_2(\omega) = \varepsilon(\omega)^{1/2}$ exhibits the same features as listed earlier for $n_2(\omega)$ and $b(\omega)$. Therefore, the dispersion relations given by eqs 5.1 and 5.2 can also be derived through the same contour integration.

The dispersion relation between the unknown phase and the measured amplitude of the reflectivity is actually a relation between the real and imaginary parts of the complex function $\ln \hat{r} = \ln(r^{1/2}) + i\theta$. In general, $\ln \hat{r}$ may contain singularities in the upper half of the complex frequency plane. In an effort to infer the behavior of $\ln \hat{r}$ from the behavior of $\hat{n}_2$, we recall Fresnel's equations for s- and p-polarization, respectively,

$$\hat{r}_\perp = \frac{n_1\cos\alpha_1 - (\hat{n}_2^2 - n_1^2\sin^2\alpha_1)^{1/2}}{n_1\cos\alpha_1 + (\hat{n}_2^2 - n_1^2\sin^2\alpha_1)^{1/2}} \qquad (5.30)$$

and

$$\hat{r}_\parallel = \frac{\hat{n}_2\cos\alpha_1 - n_1[1 - (n_1/\hat{n}_2)^2\sin^2\alpha_1]^{1/2}}{\hat{n}_2\cos\alpha_1 + n_1[1 - (n_1/\hat{n}_2)^2\sin^2\alpha_1]^{1/2}} \qquad (5.31)$$

From eqs 5.30 and 5.31, we can see that for both polarizations, $\ln \hat{r}$ may contain a singularity at the frequency $i\eta$ given by

$$n_2(i\eta)/n_1 = 1 \qquad \text{(no reflection)} \qquad (5.32)$$

or at the frequency $i\tau$ given

$$n_2(i\tau)/n_1 = \sin\alpha_1 \qquad (5.33)$$

The singularity at $i\eta$ is due to taking the logarithm of zero, whereas the one at $i\tau$ is due to taking the square root of zero. For p-polarization, the log 0 case can also occur at $i\delta$ given by

$$n_2(i\delta)/n_1 = \tan\alpha_1 \qquad (5.34)$$

which corresponds to the Brewster angle condition. Because $n_2(i\omega_2)$ decreases monotonically with ω_2, it can be seen from the above equations that $\eta < \tau$ and $\delta < \tau$. Therefore, in ATR experiments, if $\hat{n}_2(0) < n_1\sin\alpha_1$, then none of these singularities will occur in the upper imaginary axis, and $\ln \hat{r}$ becomes analytic in the upper half-plane. By setting to zero the integral of $\ln \hat{r}/(\omega - \omega')$ along the contour shown in Figure 5.1 and separating the real and imaginary parts, we can derive eq 5.5, where θ' represents the imaginary part of the limiting value of $I_D/i\pi$ and I_D is the integral along the semicircle D.

Evaluations of Kramers–Kronig Transforms

As mentioned earlier, there are two mathematically equivalent forms of the same dispersion relations. In this section we will discuss the numerical methods for evaluating the integrals involved in these dispersion relations. Suppose we need to calculate the refractive index spectrum from the following digital spectrum of the absorption index:

$$k_2(\omega_1),\ k_2(\omega_2),\ k_2(\omega_3),\ ...,\ k_2(\omega_j),\ ...,\ k_2(\omega_{m-1}),\ k_2(\omega_m)$$

If eq 5.1 is used and the integral is approximated by the summation, we obtain

$$n_2(\omega_i) = n_\infty + \frac{2}{\pi}\sum_{j=0}^{m}\left[\frac{\omega_j k_2(\omega_j)}{\omega_j^2 - \omega_i^2}\Delta\omega_j\right] \qquad (5.35)$$

The point at which $\omega_j = \omega_i$ will give an infinitely large value. To avoid such a pole in the integrand, approximate methods have to be used. One of the approaches is to calculate the sum of terms at alternate points so that the term at $j = i$ is skipped and $n_2(\omega)$ will not go to infinity. Thus, if i is even, the summation is carried out on the terms with odd indices, and if j is odd, the summation is carried out on the terms with even indices. This approach is called *Maclaurin's formula*. Another approach to avoid the pole at $j = i$ is to carry out the summation over all the points except the point at $j = i$. The result is then corrected by adding $\Delta\omega_j[k_2'(\omega_i) + k_2(\omega_i)/(2\omega_i)]/\pi$, where $k_2'(\omega_i)$ is the derivative of $k_2(\omega)$ evaluated at ω_i. The correction term represents the approximate value of the following integral

$$\frac{1}{\pi}P\int_{\omega_i-\Delta\omega_1}^{\omega_i+\Delta\omega_1}\frac{\omega k_2(\omega)}{\omega^2 - \omega_i^2}\,d\omega \qquad (5.36)$$

which is evaluated by approximating $k_2(\omega)$ with the first two terms of its Taylor series expansion with respect to $\omega - \omega_i$. This approach is called the *trapezium formula*. Depending on how $k_2'(\omega_i)$ is calculated from the discrete data, there may be several versions of the trapezium approach.

If eq 5.3 is used to calculate $n_2(\omega)$, the pole at $\omega_j = \omega_i$ does not exist. The sine and cosine transforms can be calculated by the fast Fourier transform (FFT) algorithm. Some commercial FFT routines require an input of a complex data array for both positive and negative frequency ranges. Under such circumstances, the imaginary part of the required data array can be zero-filled, and the real part of the negative frequencies can be obtained from the data given for the positive frequencies. From the behavior of the dielectric constant and its relation to the absorption index, it can be readily shown that $k_2(\omega)$ is an odd function and its sine transform is an even function. The output of the FFT algorithm is also a complex data array, whose real and imaginary parts represent the cosine and sine transforms, respectively.

Ohta and Ishida (*9*) evaluated the relative precision of various numerical Kramers–Kronig transform methods for calculating $n_2(\omega)$ from $k_2(\omega)$. The results based on a set of model functions of $n_2(\omega)$ from $k_2(\omega)$ showed that Maclaurin's formula was the most accurate approach to determining dispersive relationships. Urban and Huang (*10*) confirmed these results and further improved the error evaluation method by using the normalized root mean squares. They also evaluated the relative precision of various numerical Kramers–Kronig transform methods for calculating $k_2(\omega)$ from $n_2(\omega)$, or calculating $\theta(\omega)$ from $\ln R(\omega)^{1/2}$. Surprisingly, however, the results showed that Maclaurin's method is more accurate than the double Fourier transform method only after a baseline offset of the transformed spectra is performed. This topic is discussed further in Chapters 6 and 7.

References

1. Kramers, H. A. *Atti del Congresso Internazionale dei Fisici, 11–20 Settembre 1927*, Como–Pavia–Roma; Nicola Zanichelli: Bologna, 1928; Part 2.

2. Kronig, R. de L. *J. Opt. Soc. Am.* **1926**, *12*, 547.

3. Kronig, R. de L. *Ned. Tijdschr. Natuurk.* **1942**, *9*, 402.

4. Robinson, T. S.; Price, W. C. In *Molecular Spectroscopy*; Sell. G., Ed.; Institute of Petroleum: London, 1955.

5. Plaskett, J. S.; Schatz, P. N. *J. Chem. Phys.* **1963**, *38*, 612.

6. Bardwell, J. A.; Dignam, M. J. *J. Chem. Phys.* **1985**, *83*, 5468.

7. Toll, J. S. *Phys. Rev.* **1956**, *104*, 1760.

8. Stern, F. *Solid State Phys.* **1963**, *15*, 299.

9. Ohta, K.; Ishida, H. *Appl. Spectrosc.* **1988**, *42*, 952.

10. Urban, M. W.; Huang, J. B. *Appl. Spectrosc.* **1992**, *46*, 1666.

6

Methods of Analysis of ATR Spectra Using Kramers–Kronig Transforms

Introduction

In an ATR Fourier transform IR experiment, the measured reflectivity spectrum is a complex function of absorption (k_2) and refractive (n_2) index spectra. Because these quantities are affected by experimental conditions, to compare the results obtained under various conditions, ATR spectra should be converted to the spectra of the optical constants k_2 and n_2. If desired, the optical constants can be converted to the Beer–Lambert absorbance spectrum by using the relationship $\beta_2 = 4\pi\tilde{\nu}k_2$, in which β_2 is the absorbance per unit path length and $\tilde{\nu}$ is the wavenumber. Because the band intensities are proportional to the concentration of the absorbing species, the relationship can be used for quantitative analysis.

The early methods for determining optical constants from ATR measurements were based on the analysis of two or more ATR spectra from different ATR crystals at variable angles of incidence, using perpendicular and parallel polarizations (1–4). Because these approaches have certain limitations, ATR correction algorithms based on Kramers–Kronig transformation (KKT) (5–7) have received increased attention (8–11). Although these algorithms allow calculations of n_2 and k_2 from a single ATR spectrum of an isotropic material, the results depend on (1) what quantities are related by the transformation, (2) how KKT is carried out numerically, and (3) how KKT is incorporated into the ATR calculations. In this chapter, three ATR correction algorithms developed by Bertie and Eysel (10), Dignam and Mamiche-Afara (9), and Urban and Huang (12) will be discussed and evaluated.

3348–9/96/0083/$15.00/0/© 1996 American Chemical Society

Evaluation of Accuracy of the ATR Algorithms

Because there are no primary standards available for the optical constants, it is convenient to rely on synthesized ATR spectra. An ATR spectrum $R(\tilde{v})$ can be generated with known optical constants n_2 and k_2 and a defined set of experimental parameters.

Using such a spectrum, the ATR algorithm can be assessed by calculating the new optical constants n_2' and k_2' from the synthesized ATR spectrum obtained from the n_2 and k_2 values. Next, deviations between n_2' and n_2 and between k_2' and k_2 are evaluated to estimate the accuracy of the ATR correction algorithm. Because no experimental variables are involved in the process, this evaluation process is accurate as long as Fresnel's reflectivity theory and the causality principle remain valid.

Let us generate the model absorption spectrum using the following antisymmetric linear combinations of the Lorentzian functions (13):

$$k_2 = \frac{k_{2,\max}\chi^2}{(\tilde{v} - \tilde{v}_m)^2 + \chi^2} - \frac{k_{2,\max}\chi^2}{(\tilde{v} + \tilde{v}_m)^2 + \chi^2} \tag{6.1}$$

In the mid IR range, k_2 in eq 6.1 is dominated by the first term, which represents a Lorentzian band with $k_{2,\max}$ at $\tilde{v}_m$ and a half-width at half-maximum (HWHM) equal to χ. The presence of the second term makes it possible to calculate the corresponding refractive index spectrum using the following equation (13):

$$n_2 = n_{2,\infty} - \frac{(\tilde{v} - \tilde{v}_m)k_{2,\max}\chi}{(\tilde{v} - \tilde{v}_m)^2 + \chi^2} + \frac{(\tilde{v} + \tilde{v}_m)k_{2,\max}\chi}{(\tilde{v} + \tilde{v}_m)^2 + \chi^2} \tag{6.2}$$

where $n_{2,\infty}$ is the refractive index at infinite wavenumber. It can be shown that the optical constants generated by eq 6.1 and 6.2 satisfy the following Kramers–Kronig relation:

$$n_2(\tilde{v}) = n_\infty + \frac{2}{\pi}P\int_0^\infty \frac{sk_2(s)}{s^2 - \tilde{v}^2}ds \tag{6.3}$$

and

$$k_2(\tilde{v}) = -\frac{2\tilde{v}}{\pi}P\int_0^\infty \frac{n_2(s)}{s^2 - \tilde{v}^2}ds \tag{6.4}$$

where P is the principal value of the integral and $\tilde{v}$ and s are wavenumbers.

The ATR spectra are synthesized from the model optical constants using Fresnel's equation for perpendicular light polarization,

$$R = f(k_2, n_2)$$

$$\equiv \left| \frac{n_1 \cos \alpha_1 - [(n_2 - ik_2)^2 - n_1^2 \sin^2 \alpha_1]^{1/2}}{n_1 \cos \alpha_1 + [(n_2 - ik_2)^2 - n_1^2 \sin^2 \alpha_1]^{1/2}} \right|^2 \tag{6.5}$$

$$= \frac{\zeta^2 - Y - 2\zeta a}{\zeta^2 - Y + 2\zeta a}$$

where n_1 is the refractive index of the ATR crystal, α_1 is the angle of incidence, $\zeta = n_1 \cos \alpha_1$, $a = [(Y + X)/2]^{1/2}$, $X = n_2^2 - k_2^2 - n_1^2 \sin^2 \alpha_1$, and $Y = (X^2 + 4n_2^2 k_2^2)^{1/2}$. R is used to denote the true reflectivity data, whereas f is a quantity calculated from the n_2 and k_2 estimates using eq 6.5.

In an effort to establish how the band intensity affects the accuracy of ATR algorithms, a series of spectra can be generated in the range from 2000 to 1400 cm^{-1} using variable $k_{2,max}$ values and $\tilde{v}_m = 1700$ cm^{-1}, $\chi = 3$ cm^{-1}, $\alpha_1 = 45°$, $n_{2,\infty} = 1.5$, and $n_1 = 2.38$. The $k_{2,max}$ values may range from 0.1 to 1.5, which covers the absorption intensity changes for most organic materials. Figure 6.1 illustrates spectra generated with the model optical constants with $k_{2,max} = 0.3$ and the corresponding ATR spectrum ($-\ln R$).

In the calculation of n_2' and k_2' by iterative procedures, the iterations are completed when the deviation of $f(k_2', n_2')$ from R is minimized. The deviations between the spectra are commonly assessed as the root-mean-square differences (*13*) or simply the sum of the squared differences over all wavenumbers (*10*). To obtain the relative deviations, these approaches should incorporate an appropriate normalization. Therefore, the deviations of n_2' and k_2' are evaluated here as the normalized root-mean-square differences calculated by the following equations:

$$\delta_k = \frac{\left(\dfrac{1}{N} \sum_{i=0}^{N} [k_2(\tilde{v}_i) - k_2'(\tilde{v}_i)]^2 \right)^{1/2}}{\left(\dfrac{1}{N} \sum_{i=0}^{N} k_2^2(\tilde{v}_i) \right)^{1/2}} \tag{6.6}$$

and

$$\delta_n = \frac{\left(\dfrac{1}{N} \sum_{i=0}^{N} [n_2(\tilde{v}_i) - n_2'(\tilde{v}_i)]^2 \right)^{1/2}}{\left(\dfrac{1}{N} \sum_{i=0}^{N} [n_2(\tilde{v}_i) - n_{2,\infty}]^2 \right)^{1/2}} \tag{6.7}$$

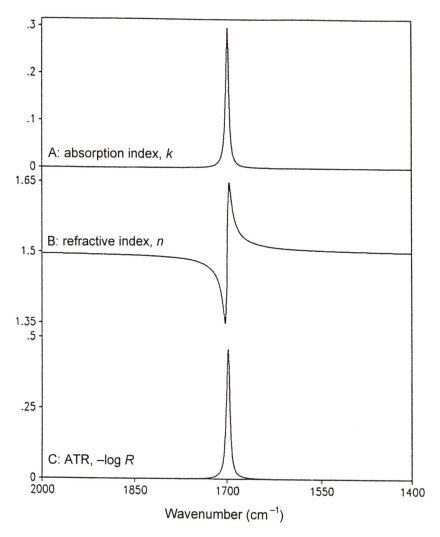

Figure 6.1. Set of model spectra with $k_{2,max} = 0.3$, used to evaluate the accuracy of the ATR algorithms: (A) absorption index spectrum, k_2; (B) refractive index spectrum, n_2; (C) ATR spectrum, $-\ln R$.

where N is the number of spectral points used in the summation. Because in quantitative analysis the primary concerns are related to the errors near the absorption band, the summations in eqs 6.6 and 6.7 were carried out in the spectral region from $\tilde{v}_m - 5\chi$ to $\tilde{v}_m + 5\chi$. This approach allows a separation of the errors in the absorbing spectral region from the errors in the nonabsorbing region—for example, a drifting baseline and the deviations near the spectral limits.

Bertie–Eysel Algorithm

Figure 6.2 is a flowchart for the Bertie–Eysel model, which involves the following steps:

1. An initial guess of the absorption index, $k_{2,i}$, is obtained using an approximate equation developed for weak bands.

2. An estimate of the refractive index, $n_{2,i}$, is calculated from the $k_{2,i}$ values, using KKT.

3. The next estimate of the absorption index, $k_{2,i+1}$, is calculated by numerically solving $R = f(k_{2,i+1}, n_{2,i})$, where f represents an appropriate Fresnel relation. For perpendicular polarization, f is given by eq 6.5.

The last two steps are iterated until a preset endpoint is reached. Step 3 in the above iteration involves an inner loop, which consists of the following steps:

3a. The $k_{2,i}$ spectrum is used as the initial guess for the solution k_{i+1} and passed to the calculation for k_2^j.

3b. The reflectivity, $R^j = f(k_2^j, n_{2,i})$, is calculated.

3c. The next estimate for the solution, k_2^{j+1}, is calculated from the following equation:

$$k_2^{j+1} = k_2^j \left(\frac{-\ln R + C}{-\ln R^j + C} \right) \tag{6.8}$$

where C is a positive constant included to avoid division by zero. The last two steps in the inner loop are repeated until a preset endpoint is reached.

Dignam–Mamiche-Afara Algorithm

Figure 6.3 is a flowchart for the ATR algorithm developed by Dignam and Mamiche-Afara (9). The algorithm is based on the following Kramers–Kronig relation between $\ln(R^{1/2})$ and the phase angle θ (Step 1) of the complex-valued Fresnel reflection coefficient $\hat{r}$ (Step 2):

$$\theta(\tilde{v}) = \theta' + \frac{2\tilde{v}}{\pi} P \int_0^\infty \frac{\ln[R(s)^{1/2}]}{s^2 - \tilde{v}^2} ds \tag{6.9}$$

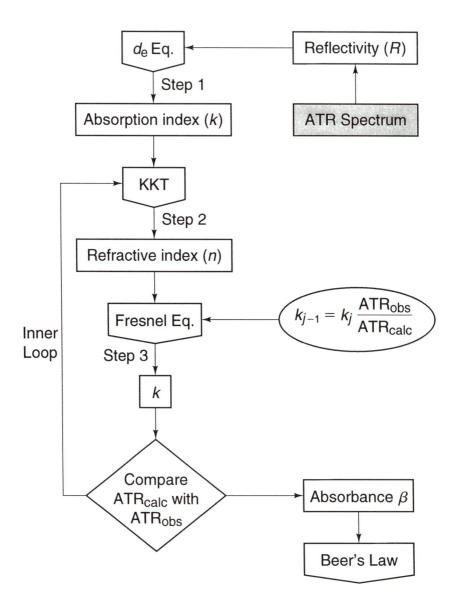

Figure 6.2. Flowchart of the Bertie–Eysel algorithm.

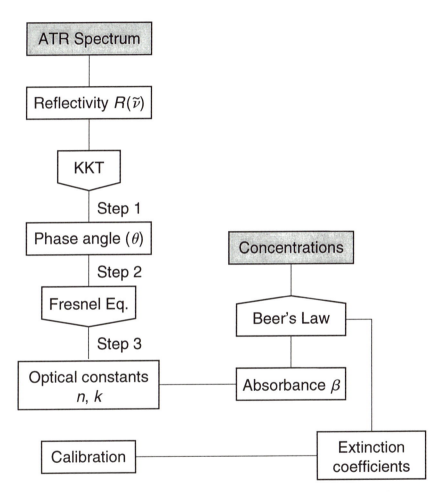

Figure 6.3. Flowchart of the Dignam–Mamiche-Afara algorithm.

where θ' is given by (16)

$$\theta' \approx -\pi + 2\arctan\frac{(n_1^2\sin^2\alpha_1 - n_{2,\infty}^2)^{1/2}}{n_1\cos\alpha_1} \tag{6.10}$$

Given the reflectivity R from the ATR measurements and the phase angle θ calculated from eq 6.9, the complex-valued Fresnel reflection coefficient $\hat{r}$ can be obtained (Step 3). This, in turn, allows direct calculation of the absorption and reflection indices from the following equations:

$$n_2 = n_1\mathrm{Re}\left[\sin^2\alpha_1 + \left(\frac{1-\hat{r}}{1+\hat{r}}\right)^2\cos^2\alpha_1\right]^{1/2} \tag{6.11}$$

and

$$k_2 = -n_1\mathrm{Im}\left[\sin^2\alpha_1 + \left(\frac{1-\hat{r}}{1+\hat{r}}\right)^2\cos^2\alpha_1\right]^{1/2} \tag{6.12}$$

n_2-to-k_2 and k_2-to-n_2 Conversions

In the algorithms developed by Bertie and Eysel (10) and by Dignam and Mamiche-Afara (9), the numerical KKT was achieved by the double Fourier transform (FT) method. However, later studies by Ohta and Ishida (13) have shown that the method based on Maclaurin's formula may be more accurate. In view of these considerations, let us first compare the Maclaurin and double FT methods when used to transform synthesized spectra from k_2 to n_2 and from n_2 to k_2.

Accuracies of the numerical KKT methods can be assessed by evaluating the deviations of the numerically calculated Kramers–Kronig n_2' and k_2' transforms. The analytical Kramers–Kronig transforms of n_2 and k_2 model spectra are given by eqs 6.1 and 6.2, respectively. Using eqs 6.5 and 6.7 and the model spectra shown in Figure 6.1, we can generate Table 6.1 which lists the δ_k and δ_n values for double FT and Maclaurin methods. Since more accurate transformation is obtained for smaller values of δ_k and δ_n, the results in Table 6.1 indicate that the Maclaurin method is more accurate for the k_2-to-n_2 transformation. Although this result is in agreement with previous studies (13), the double FT method is somewhat more accurate when used in the n_2-to-k_2 direction. Large deviations for the Maclaurin method in the n_2-to-k_2 direction result mainly from a baseline drift of the transformed spectra. In fact, with baseline offset, the Maclaurin method can be more accurate. Although these results are obtained for the model spectra generated with a $k_{2,\mathrm{max}}$ value of 0.3, it can be shown that changing $k_{2,\mathrm{max}}$ to values between 0.1 and 1.5 has no effect on the δ_k and δ_n values calculated for the double FT and Maclaurin KKT methods.

Table 6. 1. Normalized root-mean-square differences between the analytical and numerical transforms of the optical constants

Numerical method	k_2 to n_2 ($\delta_n \cdot 10^2$)	n_2 to k_2 ($\delta_k \cdot 10^2$)
Double FT	4.5	1.3
Maclaurin	0.018	1.6

Having evaluated the accuracy of the numerical KKT methods, let us establish how the ATR correction algorithms are affected by the choice of KKT method described in Chapter 2. The normalized root-mean-square errors in k_2' and n_2' spectra obtained using the ATR correction algorithms are listed in Tables 6.2 and 6.3. According to the results presented in Table 6.2, substituting the Maclaurin KKT method for the double FT method enhances the accuracy of the Dignam–Mamiche-Afara algorithm. On the other hand, the accuracy enhancement of the Bertie–Eysel algorithm is observed only for smaller $k_{2,max}$ values (less intense bands).

Table 6. 2. Normalized root-mean-square differences between k_2 and k_2' obtained by the Bertie–Eysel and Dignam–Mamiche-Afara algorithms

Numerical method	ATR algorithm	$\delta_k \cdot 10^2$				
		$k_{2,max}$=0.1	$k_{2,max}$=0.3	$k_{2,max}$=0.4	$k_{2,max}$=0.6	$k_{2,max}$=1.5
Double FT	Bertie	1.1	2.3	48.0	62.0	89.0
	Dignam	1.5	3.3	3.6	3.6	1.9
Maclaurin	Bertie	0.022	39.0	48.0	62.0	89.0
	Dignam	0.071	0.21	0.28	0.41	0.99

NOTE: The $k_{2,max}$ values indicated in this and the following tables are those used in generating the various model optical constants via eqs 6.1 and 6.2.

Table 6. 3. Normalized root-mean-square differences between n_2 and n_2' obtained by the Bertie–Eysel and Dignam–Mamiche-Afara algorithms

Numerical method	ATR algorithm	$\delta_n \cdot 10^2$				
		$k_{2,max}$=0.1	$k_{2,max}$=0.3	$k_{2,max}$=0.4	$k_{2,max}$=0.6	$k_{2,max}$=1.5
Double FT	Bertie	4.4	3.6	52.0	69.0	100.0
	Dignam	4.6	2.9	1.97	0.81	2.3
Maclaurin	Bertie	0.031	45.0	55.0	71.0	105.0
	Dignam	0.47	0.54	0.57	0.65	1.1

Urban–Huang Algorithm

The objective of the Urban–Huang algorithm is to simultaneously improve the accuracy of the algorithms for weak and strong bands. For that reason, it is first necessary to identify the origins of possible deviations. In the algorithm suitable for weak bands, the recurrence formula used in step 3 converges only if $f(k_2, n_2)$ is a monotonic, decreasing function of k_2 for a given n_2 value. However, by choosing different n_2 values, three possible curves for the Fresnel reflectivity function $f(k_2, n_2)$ can be obtained, as illustrated by plots A, B, and C of Figure 6.4. In curve A, as k_2 increases from zero and passes through a branching point $k_{2,b}$, $f(k_2, n_2)$ decreases from unity to a minimum at $k_{2,b}$ and then increases.

The branching point $k_{2,b}$ can be obtained numerically. Because $\ln[1 - f(k_2, n_2)]$ decreases monotonically with increasing $f(k_2, n_2)$, the branching point $k_{2,b}$ can be more conveniently obtained by setting $\{\partial \ln[1-f(k_2, n_2)]/\partial k_2\}_n = 0$, which gives

$$\frac{Y' + 2\zeta a'}{\zeta^2 + Y + 2\zeta a} = \frac{a'}{a} \tag{6.13}$$

where a, ζ, and Y are as defined for eq 6.5, and the prime sign (') denotes the first derivative with respect to k_2 when n_2 is a constant. Noticing that $4aa' = Y' + X'$ and $X'/Y' = Y/(X - 2n_2^2)$, we have

$$2(Y + X) = (\zeta^2 + Y)\left(1 + \frac{Y}{X - 2n_2^2}\right) \tag{6.14}$$

Numerical calculation shows that the $k_{2,b}$ value obtained from eq 6.14 decreases with increasing n_2. When n_2 is larger than a critical value $n_{2,c}$, no positive solution $k_{2,b}$ can be obtained from eq 6.14. By setting k_2 to zero and noticing that at $k_2 = 0$, $Y = -X$ when $n_2 < n_1\sin\alpha_1$, and $Y = X$ when $n_2 > n_1\sin\alpha_1$, we derive the following expression for $n_{2,c}$:

$$n_{2,c} = \frac{n_1\sin\alpha_1}{2}\left[1 + \left(1 + \frac{8}{\sin^2\alpha_1}\right)^{1/2}\right] \tag{6.15}$$

The difference between curves A and B in Figure 6.4 lies in the initial value of $f(k_2, n_2)$ at $k_2 = 0$, which is less than unity for curve B. The transition between these two patterns occurs at $n_2 = n_1\sin\alpha_1$. Curve C is characterized by the disappearance of the branching point at $k_{2,b}$; the pattern occurs when $n_2 \geq n_{2,c}$, where $n_{2,c}$ is given by eq 6.15. It is convenient to divide n_2 values into three ranges: (1) $n_2 \leq n_1\sin\alpha_1$, (2) $n_1\sin\alpha_1 < n_2 < n_{2,c}$, and (3) $n_2 \geq n_{2,c}$. Now we can see that in order for $f(k_2, n_2)$ to be a monotonic, decreasing function of k_2 so that step 3 in the Bertie–Eysel algorithm can

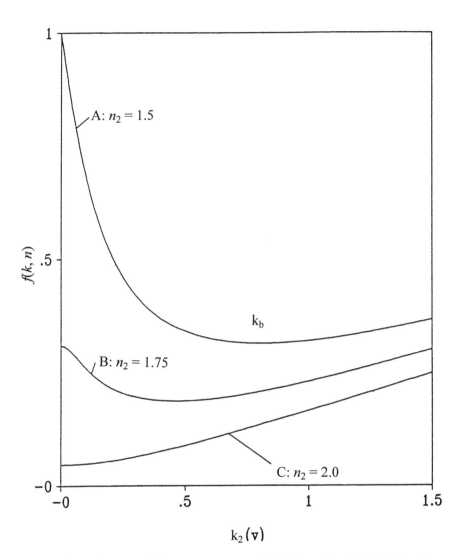

Figure 6.4. Three possible curve patterns of the Fresnel reflectivity function $f(k_2, n_2)$ *at constant* n_2: *(A)* $n_2 = 1.5$, *(B)* $n_2 = 1.75$, *and (C)* $n_2 = 2.0$.

result in finding a solution of $k_{2,i+1}$ from the equation $R = f(k_{2,i+1}, n_{2,i})$, the following conditions must be satisfied: (1) $n_{2,i}$ is in one of the first two ranges, (2) the solution $k_{2,i+1}$ and all the intermediate estimates k_2^j are less than the branching point $k_{2,b}$, and (3) at least one positive solution for $k_{2,i+1}$ exists.

Some of these conditions can be relieved through appropriate modifications. For example, conditions 1 and 2 can be replaced by a less restrictive condition 4, $\text{sign}(k_{2,i+1} - k_{2,b}) = \text{sign}(k_2^j - k_{2,b})$, if the recurrence formula in the step 3c of the Bertie–Eysel algorithm is replaced by

$$k_2^{j+1} = k_2^j \left(\frac{-\ln R + C}{-\ln R^j + C} \right)^{-\text{sign}(k_2^j - k_{2,b})} \tag{6.16}$$

Note that the new condition is necessary because the Fresnel function illustrated in Figure 6.4 may have two branches separated by $k_{2,b}$. Each of these branches may contain a solution to the equation $R = f(k_{2,i+1}, n_{2,i})$. If condition 4 is not satisfied, the iteration based on the modified recurrence formula may result in spikes in the $k_{2,i+1}$ spectrum. These spikes, in turn, may generate large errors in $n_{2,i+1}$ spectrum when KKT is performed. This problem may be overcome by including an appropriate smoothing routine in the iteration process. A failure of condition 3 is usually due to a large deviation between $n_{2,i}$ and n_2. If this occurs, no simple modifications are available.

Next, let us examine the sources of error in the algorithm suitable for strong bands. Deviations in the calculated phase shift due to the limited wavenumber range in the numerical KKT have been addressed, and correction methods for such errors have been proposed (*14, 15*). These methods, however, are inconvenient because they require refractive index values determined at several points in the spectral region of interest. Aside from the problems associated with the numerical KKT, the approximate relation given by eq 6.10 can be a source of error. Other approximations have been reported (*11*), but it has been shown that eq 6.10 is fairly accurate (*16*).

Based on these considerations, Figure 6.5 is a flowchart for the Urban–Huang algorithm that incorporates the use of the exact Kramers–Kronig relation between n_2 and k_2 to refine the initial results obtained by the following steps:

1. An initial guess of the $k_{2,i}$ values is generated by the original Dignam–Mamiche-Afara (*9*) algorithm, and used to obtain the initial guess for $n_{2,i}$ using KKT.

2. An approximate phase shift θ_i is calculated from the $n_{2,i}$ and $k_{2,i}$ values using the following relationship:

$$\theta_i(\tilde{v}) = \arctan\left(\frac{2\zeta b_1}{Y_i - \zeta^2} \right) \tag{6.17}$$

where $b_i = [(Y_i - X_i)/2]^{1/2}$, $Y_i = (X_i^2 + 4n_{2,i}^2 k_{2,i}^2)^{1/2}$, and $X_i = n_{2,i}^2 - k_{2,i}^2 - n_1^2 \sin^2\alpha_1$. Equation 6.17 is derived from Fresnel's equation, given by eq 6.5.

3. The next estimate is obtained by calculating k_2 from eq 6.12 using $\hat{r}_i = R^{1/2}\exp(i\theta_i)$. This is used as the estimate for the true complex

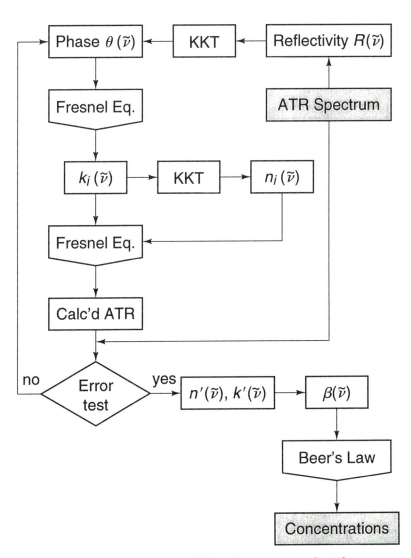

Figure 6.5. Flowchart of the Urban–Huang algorithm.

Fresnel reflection coefficient $\hat{r}$, and the result is passed to the calculation for $k_{2,i+1}$.

4. The next estimate of n_2 is obtained from $k_{2,i+1}$ by KKT, and the result is passed to the calculation for $n_{2,i+1}$.

As illustrated in Figure 6.5, steps 2 through 4 are repeated until the preset endpoint is reached. The normalized root-mean-square errors in k_2' and n_2' obtained with this algorithm are listed in Table 6.4. Comparing them with the results in Tables 6.2 and 6.3 reveals that when the Maclaurin method is used for the numerical KKT, the new algorithm exhibits better accuracy for a wider range of $k_{2,max}$ values. Despite the significant improvements, two issues concerning this approach should be addressed. The first is the accuracy of the numerical KKT. When the double FT method is used for numerical KKT, the Urban–Huang algorithm is effective as long as weak and moderate-intensity bands are used but fails for strong bands. Second, the initial guesses for $n_{2,i}$ and $k_{2,i}$ must be reasonably good. For example, when used to refine the initial estimates employed in the algorithm for weak bands, the iterative routine in the new algorithm shows little improvement over the Bertie–Eysel algorithm when strong bands are tested.

Table 6.4. Normalized root-mean-square differences for k_2' and n_2' obtained by the Urban–Huang algorithm

Numerical method		$k_{2,max}=0.1$	$k_{2,max}=0.3$	$k_{2,max}=0.4$	$k_{2,max}=0.6$	$k_{2,max}=1.5$
Double FT	$\delta_k \cdot 10^2$	1.1	2.4	2.6	3.6	1.9
	$\delta_n \cdot 10^2$	4.3	3.5	3.0	7.6	5.8
Maclaurin	$\delta_k \cdot 10^2$	0.0052	0.025	0.055	0.26	0.98
	$\delta_n \cdot 10^2$	0.019	0.027	0.060	0.30	0.99

Examples of Applications

Figure 6.6 shows an ATR spectrum of poly(ethyl acrylate) (PEAc) in the 3500–2500 and 2000–500 cm^{-1} regions obtained using the parameters listed in Table 6.4 and corrected for multiple reflections. Because the bands in the 3500–2500 cm^{-1} region are relatively weak, they will be used to illustrate the application of the ATR algorithms for weak bands, whereas the 2000–500 cm^{-1} region will serve for strong bands. For each region, the absorption index spectrum k_2' will be calculated by using the ATR algorithms in conjunction with the Maclaurin KKT method, from which the ATR spectrum in the form of $-\ln[f(k_2', n_2')]$ can be calculated (n_2' is the refractive index spectrum calculated from k_2' by KKT). Figure 6.7 illustrates ATR spectra in the strongly absorbing C=O stretching region. Trace

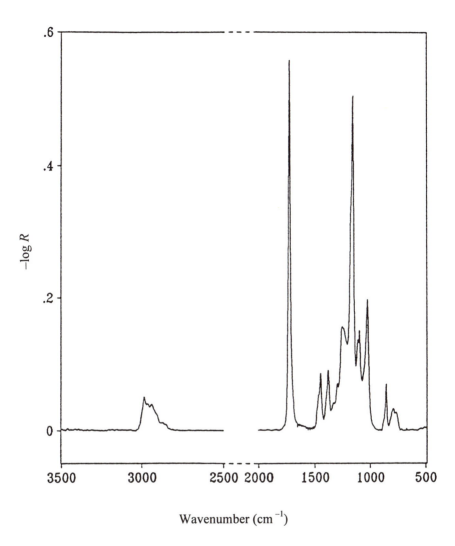

Figure 6.6. ATR FT-IR spectrum of poly(ethyl acrylate). (Reproduced with permission from reference 12. Copyright 1992 Society for Applied Spectroscopy.)

A is the measured single-reflection ATR spectrum, and traces B, C, and D are the ATR spectra calculated from the k_2' spectra using Bertie–Eysel, Dignam–Mamiche-Afara, and Urban–Huang, respectively. The corresponding ATR spectra in the weakly absorbing C–H stretching region are illustrated in Figure 6.8. A comparison of the results presented in Figures 6.7, 6.8, and 6.9 demonstrates that the algorithm proposed in Figure 6.5 provides the best fit.

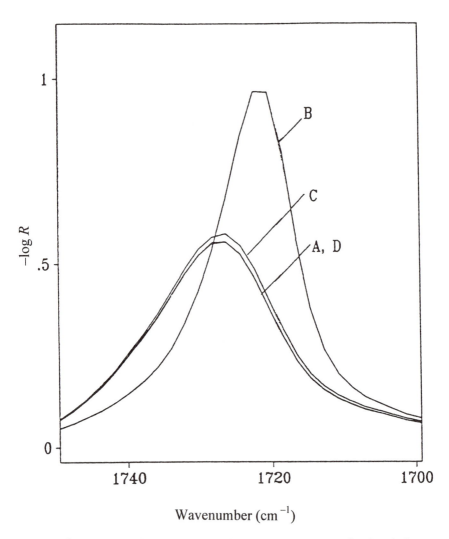

Figure 6.7. Measured and calculated ATR FT-IR spectra of poly(ethyl acrylate) in the strongly absorbing C=O stretching region: (A) measured, (B) Bertie–Eysel, (C) Dignam–Mamiche-Afara, and (D) Urban–Huang. (Reproduced with permission from reference 12. Copyright 1992 Society for Applied Spectroscopy.)

Another illustrative example is the C=O stretching bands of an acrylic isocyanate coating. The original ATR spectrum was measured using the experimental parameters listed in Table 6.5, and the C=O stretching region is shown in Figure 6.9.

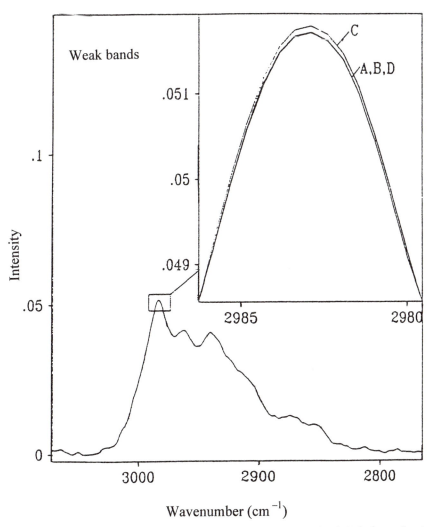

Figure 6.8. Measured and calculated ATR FT-IR spectra of poly(ethyl acrylate) in the weakly absorbing C–H stretching region: (A) measured, (B) Bertie–Eysel, (C) Dignam–Mamiche-Afara, and (D) Urban–Huang. (Reproduced with permission from reference 12. Copyright 1992 Society for Applied Spectroscopy.)

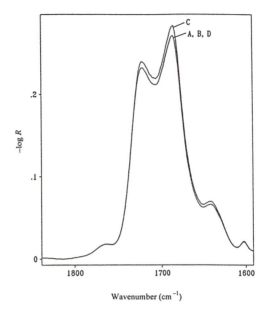

Figure 6.9. Measured and calculated ATR FT-IR spectra of an acrylic isocyanate coating in the C=O stretching region: (A) measured, (B) Bertie–Eysel, (C) Dignam–Mamiche-Afara, and (D) Urban–Huang. (Reproduced with permission from reference 12. Copyright 1992 Society for Applied Spectroscopy.)

Table 6.5 Parameters used in ATR measurements

	Material	Poly(ethyl acrylate)	Acrylic isocyanate coating
Sample	Refractive index (589 nm)	1.4685	1.5314
	Length	20 mm	21 mm
	Thickness	100 μm	50 μm
	Position	Centered on one side of crystal	
ATR crystal	Material	KRS-5	
	Refractive index	2.4	
	Dimension	$50 \times 20 \times 3$ mm^3	
	End bevel	45°	
Incident light	Angle of incidence in air	60°	45°
	Angle of incidence in crystal	51.2°	45°
	Polarization	Perpendicular	
Data acquisition	Resolution	4 cm^{-1}	
	Number of scans	200	

Experimental Procedures Leading to Determination of Optical Constants

In Chapter 1, we established that when light propagates through an interface, a fraction may be reflected or transmitted, depending on the angle of incidence and the refractive properties of the two substances forming the interface. In transmission measurements (that is, when the incident light is normal to the surface), the propagating waves complicate the spectrum through reflection losses and multiple reflections (*17*). As discussed earlier, such problems can be corrected by knowing the optical constants, and methods for doing so have been outlined. As a matter of fact, it is always a good practice to determine optical constants as a function of wavenumber, whether the purpose is to determine absorption and refractive index spectra themselves, to perform quantitative analysis of a mixture of compounds, or just to identify normal vibrations by the band position and intensity. This is because the refractive, and therefore absorption index spectra, often undergo major changes, especially if the extinction coefficient changes rapidly. There are tables of refractive index data reported for selected compounds in the literature, collected over long periods (*18*).

As pointed out by Hawranek and Jones (*17*), determination of the true absorption spectrum from transmission measurements of liquids is complicated by reflection losses and multiple reflections at the interfaces. To compensate for these optical effects, it is necessary to know the optical constants for a liquid sample of interest. Such optical constants can be determined by a procedure that involves obtaining the absorption index spectrum from the raw absorption spectrum and then performing KKT on the absorption index spectrum to get the refractive index spectrum (*19, 20*). The values of the complex refractive index can be used to compensate for reflection losses and other optical properties. This method has led to successful analysis of the optical constants of many liquids, but the main source of error was the difficulty of determining the cell thickness. For that reason, Bertie and Eysel (*10*) used a cylindrical, multiple-reflection cell (Circle, SpectraTech), along with an iterative procedure similar to that of Hawranek and Jones, to determine the optical constants. This procedure requires a calibration liquid with known absorption and refractive index spectra, from which the effective number of reflections can be determined at a given angle of incidence. In an effort to avoid multiple reflections, Bardwell and Dignam (*21*) used a single-reflection method to determine optical constants of organic liquids. This procedure involved KKT on the logarithm of the reflectance for the s-polarized electric vector normal to the plane of incidence (*16*). Whereas these studies were concerned with isotropic media such as liquids, Dignam and Mamiche-Afara (*9*) extended the approach to anisotropic media. Finally, the Urban–Huang algorithm eliminates many of the problems associated with the previous algorithms. In any approach, usually KKT is used to obtain optical constants from spectra, providing that the refractive index of a sample of interest in a non-absorbing region is known. This usually requires an independent experiment in which the refractive index of the sample is measured in a nonabsorbing region of the visible part of the spectrum, and such measurements are extrapolated to infrared. This approach, however, has several limitations; for example, it cannot be applied to opaque or highly scattering samples in the visible region. Tickanen et al. (*22*) developed an approach to overcome these limitations by a concurrent determination of the anchor point and the optical constants using only variable-angle ATR measurements and the subtractive form obtained from KKT. The method involves ratioing the negative log ATR band intensities of weak bands recorded at different angles of incidence and an iterative computation of n_{21} ($n_{21} = n_2/n_1$) values using a fairly complicated function $f(n_{21})$. The value for the anchor point (that is, n_2 in a nonabsorbing region) is determined by averaging individual values of n_2 over the wavenumber range spanned by a weak negative log ATR intensity. Apparently, this approach requires non-polarized radiation and a sample thicker than the sampling depth of the IR radiation.

Summary

In this chapter, accuracies of commonly used ATR calculation algorithms (*9, 10, 12*) were assessed. Evidently, the approach proposed by Bertie and Eysel, intended to determine the optical constants by directly solving two independent equations based on the Fresnel reflectivity theory and the Kramers–Kronig relation between n_2 and k_2, is applicable for weak bands and fails for other cases. Large deviations for strong bands are caused by the nonmonotonic relationship between the Fresnel reflectivity and the absorption index values, which leads to divergence during iteration. Significant errors may also occur when the initial estimates of n_2 are far from the actual values, so that no positive values of k_2 can be fitted to the equation $R = f(k_2, n_2)$. On the other hand, the Dignam–Mamiche-Afara approach, involving calculations of the phase angle θ from reflectivity data by Kramers–Kronig correlation, is applicable in the entire range of absorption intensity tested in this study. This approach, however, is less accurate for weak bands. After the analysis of the possible deviation sources for weak- and strong-band algorithms, Urban–Huang ATR correction algorithm was outlined. In this algorithm, the absorption index spectrum obtained from a modification of the Dignam–Mamiche-Afara theory can be refined by an iterative process that minimizes the difference between the true and calculated reflectivity spectra while maintaining the exact Kramers–Kronig relation between n_2 and k_2. Furthermore, when the Maclaurin method for numerical KKT is used in conjunction with this algorithm, a significant improvement over all numerical calculations can be achieved.

References

1. Fahrenfort, J.; Visser, W. M. *Spectrochim. Acta* **1962**, *18*, 1103.

2. Hirschfeld, T. *Appl. Spectrosc.* **1970**, *24*, 277.

3. Crawford, B., Jr.; Goplen, T. G.; Swanson, D. In *Advances in Infrared and Raman Spectroscopy*; Clark, R. J. H.; Hester, R. E., Eds.; Heyden: London, 1980; Vol. 4, Chapter 2.

4. Hansen, W. N. *Spectrochim. Acta* **1965**, *21*, 815.

5. Frohlish, H. *Theory of Dielectrics*; Clarendon: Oxford, 1949.

6. Maeda, S.; Schatz, P. N. *J. Chem. Phys.* **1962**, *34*, 571.

7. Hansen, W. N.; Abdou, W. A. *J. Opt. Soc. Am.* **1977**, *67*, 1537.

8. Tshmel, A. E.; Vettegren, V. I. *Spectrochim. Acta* **1973**, *29A*, 1681.

9. Dignam, M. J.; Mamiche-Afara, S. *Spectrochim. Acta* **1988**, *44A*, 1435.

10. Bertie, J. E.; Eysel, H. H. *Appl. Spectrosc.* **1985**, *39*, 392.

11. Plaskett, J. S.; Schatz, P. N. *J. Chem. Phys.* **1963**, *38*, 612.

12. Urban, M. W.; Huang, J. B. *Appl. Spectrosc.* **1992**, *46*, 1666.

13. Ohta, K.; Ishida, H. *Appl. Spectrosc.* **1988**, *42*, 952.

14. Roessler, D. M. *Br. J. Appl. Phys.* **1965**, *16*, 1119.

15. Roessler, D. M. *Br. J. Appl. Phys.* **1966**, *17*, 1313.

16. Bardwell, J. A.; Dignam, M. J. *J. Chem. Phys.* **1985**, *83*, 5468.

17. Hawranek, J. P.; Jones, R. N. *Spectrochim. Acta* **1976**, *32A*, 111.

18. Bertie, J. E.; Zhang, S. L.; Eysel, H. H.; Baluja, S.; Ahmed, M. K. *Appl. Spectrosc.* **1993**, *47*, 1100.

19. Goplen, T. G.; Cameron, D. G.; Jones, R. N. *Appl. Spectrosc.* **1980**, *34*, 657.

20. Cardonna, M. In *Optical Properties of Solids*; Nudelman, S.; Mitra, S. S., Eds. Plenum: New York, 1969.

21. Bardwell, J. A.; Dignam, M. J. *Anal. Chim. Acta* **1986**, *181*, 253.

22. Tickanen, L. D.; Tejedor-Tejedor, M. I.; Anderson, M. A. *Appl. Spectrosc.* **1992**, *46*, 1846.

7

Quantitative Surface Depth Profiling

Introduction

This chapter focuses on two types of samples in ATR spectroscopy: homogeneous and nonhomogeneous samples. As indicated earlier, there are numerous examples of using ATR FT-IR spectroscopy for surface depth profiling (*1–4*). Most studies, however, used the well-established reflection theory (*5*) for homogeneous samples. As discussed in Chapter 3, this theory predicts that the penetration depth d_p can be controlled by changing the refractive index of the ATR crystal and the angle of incidence. Although this approach is indeed valid and highly useful for homogeneous systems, the quantitative relations must be modified for analysis of nonhomogeneous surfaces. This issue is particularly relevant in situations with varying surface concentrations because in most sampling situations the surfaces are not homogeneous.

Although reflection theory has been known for a long time, it was not applied to stepwise stratified media until recently (*6*). Such an approach can establish the relationship between the reflectivity data obtained from ATR FT-IR surface depth profiling experiments and the concentration at a given surface depth. In spite of the mathematical complexity of the process, an unknown surface depth profiling can be calculated by using a linear interpolation or by applying a function with variable parameters. Because the assumptions are limited and the computation time is reasonable, even when the sample is finely divided to achieve high spatial resolution, the linear interpolation approach seems to be particularly advantageous. This chapter will test the proposed method for determining the distribution of small molecules near a polymer surface and explain the calculation of the surface depth profiling. But first let us examine the drawbacks that arise from using ATR spectroscopy without taking into account optical effects.

3348–9/96/0105/$16.50/0/© 1996 American Chemical Society

Non-KKT Analysis

One approach is to treat ATR FT-IR spectra as if they were recorded in the transmission mode of detection. In an effort to use such ATR measurements in quantitative analysis, the Beer–Lambert law can be used. From the intensity measurements, the concentration of a species of interest can be determined if the extinction coefficient of a characteristic band is known. The relationship can be expressed by the following equation:

$$-\ln R(\tilde{v}) = \ln\left(\frac{I_0}{I_R}\right) = \beta(\tilde{v})d_e(\tilde{v}) \tag{7.1}$$

where I_0 and I_R are the incident and reflected light intensities, $\beta(\tilde{v})$ is the absorbance per unit length, and $d_e(\tilde{v})$ is the effective thickness. Because even for isotropic samples the effective thickness depends on the polarization of the incident light, polarized light should be used to simplify the expression for the effective thickness. The effective thicknesses for perpendicular and parallel polarizations are given by the following equations (7):

$$d_{e,\perp} = \frac{n_{21}\lambda_1\cos\theta}{\pi(1 - n_{21}^2)(\sin^2\theta - n_{21}^2)^{1/2}} \tag{7.2}$$

and

$$d_{e,\parallel} = \frac{n_{21}\lambda_1(2\sin^2\theta - n_{21}^2)\cos\theta}{\pi(1 - n_{21}^2)(1 + n_{21}^2)(\sin^2\theta - n_{21}^2)^{3/2}} \tag{7.3}$$

Equations 7.2 and 7.3 were derived for weak bands for which the refractive index of the sample can be assumed to be invariant with the wavelength of the incident light. As seen from eqs 7.1 through 7.3, under such conditions the ATR spectrum multiplied by wavenumber differs from the true absorbance spectrum $\beta(\tilde{v})$ only by a scaling factor. In this case, spectral distortions, such as frequency shifts and intensity changes, are not taken into account because for weakly absorbing bands these effects may be negligible. Therefore, conventional techniques for quantitative analysis, such as band ratioing, normalization against an internal standard band, and spectral subtractions, are directly used (8) as long as the bands are weak. Several successful applications of these data-processing techniques have been reported (9–11). However, considerable frequency shifts and intensity changes in ATR spectra relative to the true absorbance spectrum have been reported for strong bands (12, 13). Therefore, previous efforts in interpretation of ATR spectra require additional considerations. With these limitations in mind, and considering that specimen nonhomogeneity needs to be taken into account, let us establish theoretical foundations for ATR analysis.

Origin of Nonhomogeneous Approaches

The first theoretical treatment of reflectivity data for surface depth profiling was proposed by Tompkins (*14*). By introducing nonhomogeneity into Harrick's reflection theory for homogeneous surfaces (*5*), Tompkins derived the expression for the total amount of absorbed light energy at each wavenumber for a single reflection:

$$A = \frac{n_2 E_0^2}{n_1 \cos\alpha_1} \int_0^\infty \beta(z) e^{-2z/d_p} dz \tag{7.4}$$

where n_1 and n_2 are the refractive index values of the ATR crystal and the sample, respectively; α_1 is the angle of incidence; β is the absorbance per unit length; E_0 is the amplitude of the electric field at the crystal–sample interface; z is depth in the sample; and d_p is the penetration depth, given by

$$d_p = \frac{\lambda}{2\pi(n_1^2 \sin^2\alpha_1 - n_2^2)^{1/2}} \tag{7.5}$$

where λ is the wavelength of the electromagnetic radiation in vacuum (*15*). Eliminating the term $n_2 E_0^2/(n_1 \cos\alpha_1)$ in eq 7.4 by using an internal reference, and treating n_2 as a real constant in the penetration depth calculations, Tompkins outlined the method for calculating the parameters for a given assumed depth profile of β. In this approach, β may decrease or increase linearly or in steps. This method was demonstrated by analyzing the reflectivity data reported by Carlsson and Wiles (*16*), who examined polypropylene surfaces oxidized by O_2 corona discharge and analyzed their own data using the effective thickness equations for thin films (*17*). Harada et al. (*18*) applied Tompkins's method to characterize the results of saponification in depth profiling studies of a photographic film. To avoid guessing a functional form for the unknown depth profile, Hirschfeld (*19*) applied an inverse Laplace transform to solve an integral equation similar to eq 7.4, but such an approach may be limited because of the difficulty of numerical inversion of Laplace transforms because of the "ill-conditioned problem" (*20*). Small deviations in the experimental data and the errors resulting from the rounding of intermediate results during calculations can cause deviations in the results, and various numerical techniques have been proposed to overcome such problems (*21*).

Both Tompkins and Hirschfeld assumed that the evanescent waves are not perturbed by the surface absorption process. But as pointed out by Fina and Chen (*22*), the validity of this assumption holds only for weak bands, and because eq 7.4 is the Laplace transform of the absorption coefficient, it transforms angular space into distance space. Perhaps the advantage of using this approach is that when the absorption coefficient is low (for weak bands), eq. 7.4 can be solved using real rather than complex variables, but its applicability to nonhomogeneous surfaces is limited. Although for homogeneous samples the perturbation of the evanescent wave due to light absorption can be corrected by replacing eq 7.5 with a more general equation derived by Hansen (23), such an equation is not available for nonhomogeneous samples because the evanescent waves no longer follow an exponential decay. A more advanced theory was developed by Stuchebryukov (24), but in spite of its complexity, this theory is still limited to thin films with weak absorptions. Even though several model studies were conducted (22), Fina correctly pointed out that this approach is limited to bands with a small absorption coefficient (25).

Since the validity of eq 7.1 is the primary source of limitations for the existing depth profiling theories, a more rigorous expression for ATR band intensities for nonhomogeneous surfaces is needed. Such an expression can be obtained by numerically slicing a sample to form a stack of parallel thin homogeneous films (26). By applying well-established reflectivity optical theory for each layer and stepwise treatments (22, 27), we can obtain the distribution of species for nonhomogeneous surfaces.

If this approach can be successfully applied, it may have many practical applications. For example, recently, Urban and co-workers investigated surfactant distribution in latex films (28–39). This issue is particularly important because the macroscopic properties of polymeric films, such as adhesion and degradation, are greatly affected by concentration changes across the film. For example, it was found that excessive concentrations of surfactants on the film–air and film–substrate interfaces may be affected by the substrate surface tension, surfactant and copolymer structures, and mechanical deformations (30–31). Although in these studies the issue of molecular-level interactions between surfactants and copolymers was addressed for the first time, further understanding of such interactions may be obtained by quantitative assessment of the surfactant depth profile in latex films. With these considerations in mind, this chapter will outline an analytical design of the surface slicing in an effort to establish a set of equations that, in turn, will allow a correlation of the measurable reflectivity and the unknown surface depth profile. The essence of the method is to slice the surface into layers of equal thickness. Each layer is homogeneous within itself, but the layers differ from one another; each layer may have a different homogeneous distribution of species.

Theoretical Considerations of Surface Slicing

Figure 7.1 illustrates reflection conditions at the boundary between an ATR crystal and a stratified sample containing N layers with parallel boundaries

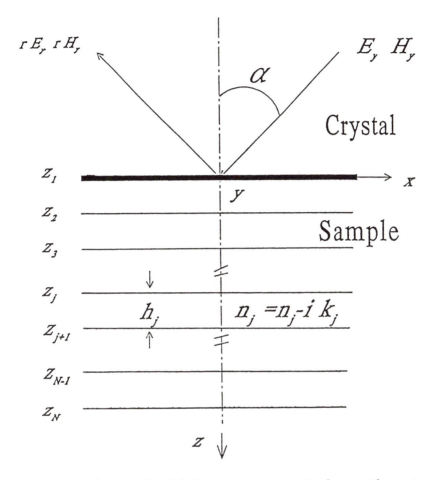

Figure 7.1. Reflection of a light beam propagating in the x–z plane at a boundary (x–y plane) between an ATR crystal (j = 0) and an N-phase strat-ified sample (j = 1, 2, ..., N); α_1, angle of incidence; $n_{2,j}$, refractive index; $k_{2,j}$, absorption index; h_j, layer thickness ($h_0 = h_N = \infty$); E_y, y component of the electric field; H_y, y component of the magnetic field of incident light; $r_\perp$, reflectivity for perpendicular polarization; $r_\parallel$, reflectivity for parallel polar-ization.

at $z = z_1, z_2, ..., z_N$. At each boundary, z_j, and within the layer below, the response of the sample to local evanescent waves can be characterized by a complex refractive index defined by $\hat{n}_{2,j} = n_{2,j} - ik_{2,j}$, where the imagi-nary part $k_{2,j}$ is referred to as the absorption index. With the notation given in Figure 7.1, the reflectivity $\hat{r}$ for perpendicular light polarization, defined as the ratio of the electric fields of the reflected and incident light beams at a wavenumber $\tilde{v}$, is given by

$$\hat{r} = R^{1/2}e^{i\theta} = \frac{(m_{11} + m_{12}P_N)P_0 - (m_{21} + m_{22}P_N)}{(m_{11} + m_{12}P_N)P_0 + (m_{21} + m_{22}P_N)} \qquad (7.6)$$

where m_{11}, m_{12}, m_{21}, and m_{22} are the elements of a 2×2 matrix (22), given by

$$\begin{bmatrix} m_{11} & m_{12} \\ m_{21} & m_{22} \end{bmatrix} = \prod_{j=1}^{N-1} \begin{bmatrix} \cos\eta_j & \frac{-i}{\xi_j}\sin\eta_j \\ -i\xi_j\sin\eta_j & \cos\eta_j \end{bmatrix} \qquad (7.7)$$

$$\eta_j = 2\pi\tilde{v}(z_{j+1} - z_j)(\hat{n}_{2,j}^2 - n_1^2\sin^2\alpha_1)^{1/2} \qquad (7.8)$$

and

$$\xi_{j,\perp} = (\hat{n}_{2,j}^2 - n_1^2\sin^2\alpha_1)^{1/2} \qquad (7.9)$$

For parallel polarization, the reflectivity $\hat{r}$ is defined as the ratio of the magnetic fields of the reflected and incident light beams.* With this definition, eqs 7.6 through 7.8 remain valid when eq 7.9 is replaced by

$$\xi_{j,\|} = \frac{(\hat{n}_{2,j}^2 - n_1^2\sin^2\alpha_1)^{1/2}}{\hat{n}_{2,j}^2} \qquad (7.10)$$

In the depth profiling experiments it is desirable to calculate the composition of each layer from the absorption index $k_{2,j}$. If the reflectivity R is measured at a wavenumber at which only the component of interest absorbs light, the concentration c_j of that component in each layer can be related to the absorption index $k_{2,j}$ by the following relation (40):

$$k_{2,j} = \frac{1}{4\pi\tilde{v}}c_j\varepsilon \qquad (7.11)$$

where ε is the extinction coefficient. Replacing molar concentrations c_j with volume fractions ϕ_j, and assuming no excess volume of mixing, eq 7.11 can be rewritten as

$$k_{2,j} = \phi_j k_2^\circ \qquad (7.12)$$

where k_2° is the absorption index of the absorbing component in its pure state. The Kramers–Kronig relation between $n_{2,j}$ and $k_{2,j}$ (41–43) leads to the refractive index $n_{2,j}$ expressed as

$$n_{2,j} = n_{2,\infty} + \phi_j\Delta n_2^\circ \qquad (7.13)$$

---

*The quantity actually measured is $R = |\hat{r}|^2$.

where Δn_2° is the Kramers–Kronig transform of k_2° and $n_{2,\infty}$ is the refractive index at infinite wavenumber. The value of $n_{2,\infty}$ will be assumed to be the same for all layers of the sample. The k_2° and n_2° values can be determined from transmission or ATR measurements (44–46).

According to eqs 7.6 through 7.13, the experimentally determined reflectivity R can be correlated with the volume fraction in each layer. Such a correlation can be represented vectorially as $R = F(\mathbf{U}, \mathbf{V}, \mathbf{W})$, where $\mathbf{U} = [n_1, \alpha_1]$, $\mathbf{V} = [\phi_1(z_1), \phi_2(z_2), ..., \phi_N(z_N)]$, and $\mathbf{W} = [n_{2,\infty}, \Delta n_2^\circ, \Delta k_2^\circ, \tilde{v}]$. The vector $\mathbf{U}$ contains all experimentally controllable parameters, $\mathbf{W}$ includes the optical constants of the pure components at a given wavenumber $\tilde{v}$, and $\mathbf{V}$ represents the unknown depth profile. Thus, given a series of reflectivity values $R_1, R_2, ..., R_K$, calculation of the depth profile $\mathbf{V}$ is reduced to the following set of nonlinear equations:

$$R_1 = F(\mathbf{U}_1, \mathbf{V}, \mathbf{W})$$
$$R_2 = F(\mathbf{U}_2, \mathbf{V}, \mathbf{W})$$
$$\vdots \tag{7.14}$$
$$R_K = F(\mathbf{U}_K, \mathbf{V}, \mathbf{W})$$

In this, as well as many IR quantitative analyses, in order to compensate for experimental conditions, an internal reference band is required for normalization. If we use the normalized intensities $Y_i = (-\ln R_i)/(-\ln R_i^{\text{ref}})$, eq 7.14 can be written as

$$Y_1 = [-\ln F(\mathbf{U}_1, \mathbf{V}, \mathbf{W})]/[-\ln F(\mathbf{U}_1, \mathbf{V}^{\text{ref}}, \mathbf{W}^{\text{ref}})]$$
$$Y_2 = [-\ln F(\mathbf{U}_2, \mathbf{V}, \mathbf{W})]/[-\ln F(\mathbf{U}_2, \mathbf{V}^{\text{ref}}, \mathbf{W}^{\text{ref}})]$$
$$\vdots \tag{7.15}$$
$$Y_K = [-\ln F(\mathbf{U}_K, \mathbf{V}, \mathbf{W})]/[-\ln F(\mathbf{U}_K, \mathbf{V}^{\text{ref}}, \mathbf{W}^{\text{ref}})]$$

In eq 7.15, the number of variables can be reduced by an appropriate assumption for the depth profile of the reference component $\mathbf{V}^{\text{ref}}$. For example, we can choose a reference component that is uniformly distributed in the sample (14). If the sample is a binary mixture, the volume fractions of the two components are related by $\phi_j^{\text{ref}} = 1 - \phi_j$. For specimens containing more than two components, we may assume that each component has the same depth profile. With this assumption, the components of the sample can be classified into two categories according to their depth profiles, and the total volume fraction of each category is related by $\phi_j^{\text{ref}} = 1 - \phi_j$.

In many sampling situations, no isolated band arising from the component of interest is present. Under such circumstances, if we can find a band in which only the component of interest and the reference component absorb IR light, eq. 7.15 can still be used when eqs 7.12 and 7.13 are replaced with the following equations:

$$\begin{bmatrix} k_{2,j}(\tilde{v}) \\ k_{2,j}(\tilde{v}^{\text{ref}}) \end{bmatrix} = \begin{bmatrix} k_2^{\circ}(\tilde{v}) & k_2^{\circ\text{ref}}(\tilde{v}) \\ k_2^{\circ}(\tilde{v}^{\text{ref}}) & k_2^{\circ\text{ref}}(\tilde{v}^{\text{ref}}) \end{bmatrix} \begin{bmatrix} \phi_j \\ \phi_j^{\text{ref}} \end{bmatrix} \tag{7.16}$$

and

$$\begin{bmatrix} n_{2,j}(\tilde{v}) - n_{\infty} \\ n_{2,j}(\tilde{v}^{\text{ref}}) - n_{\infty} \end{bmatrix} = \begin{bmatrix} \Delta n_2^{\circ}(\tilde{v}) & \Delta n_2^{\circ\text{ref}}(\tilde{v}) \\ \Delta n_2^{\circ}(\tilde{v}^{\text{ref}}) & \Delta n_2^{\circ\text{ref}}(\tilde{v}^{\text{ref}}) \end{bmatrix} \begin{bmatrix} \phi_j \\ \phi_j^{\text{ref}} \end{bmatrix} \tag{7.17}$$

Although eqs 7.16 and 7.17 may be generalized to the case where more than two components are absorbing at the same wavenumber, the large number of unknowns in the combined equation set makes it difficult to determine the depth profiles.

Numerical and Analytical Methods

Having identified the relationships that correlate the reflectivity data with the unknown depth profile of the sample, let us establish the criteria for solving these equations. For eqs 7.14 and 7.15, which are nonlinear, perhaps the most convincing way is to try out many possible depth profiles to minimize the deviation of the calculated reflectivity values from the measured data. Depending on whether the absolute (eq 7.14) or normalized (eq 7.15) reflectivity values are chosen, the deviation of the calculated reflectivity values can be calculated as the sum of the residual squares given by

$$Q = \sum_{i=1}^{K} [R_i - F(\mathbf{U}_i, \mathbf{V}, \mathbf{W})]^2 \tag{7.18}$$

or

$$Q = \sum_{i=1}^{K} \left[Y_i - \left(\frac{-\ln F(\mathbf{U}_i, \mathbf{V}, \mathbf{W})}{-\ln F(\mathbf{U}_i, \mathbf{V}^{\text{ref}}, \mathbf{W}^{\text{ref}})} \right) \right]^2 \tag{7.19}$$

If the volume fraction ϕ_j of each layer is treated as an independent variable, such a trial-and-error process is laborious. This is because in order for the stepwise stratification model to approach the actual depth profile, the sample should be divided into a large number of thin layers, resulting in a larger number of unknowns in eqs 7.14 and 7.15. Therefore, we will use two fairly straightforward approaches: linear interpolation and the use of a function with variable parameters as a trial depth profile. In the first approach, the sample surface will be divided into a small number of primary layers near the surface, and these primary layers will be further divided into thinner layers. The concentrations at the boundaries of the primary layers are independently varied during the trial-and-error process, whereas the concentrations at the sublayer boundaries are linearly interpolated. In the second approach, the volume fraction of each layer is calculated from an assumed function with a limited number of parameters, whose values are determined by minimizing the sum of the residual squares in eqs 7.18 and 7.19.

During the trial-and-error process, although the Q values calculated by eqs 7.18 and 7.19 can be used as the criteria for evaluating fit, it is also helpful to graphically illustrate the difference between the measured and calculated reflectivity data. For this reason, we may first convert the experimentally determined vector $\mathbf{U}$ into the penetration depth (d_p) using eq 7.5, along with the plot of reflectivity data as a function of d_p. The problem with this approach, however, is that the reflectivity values, or their ratios, vary with d_p even for a highly homogeneous surface. To highlight the effect of nonhomogeneity, we will convert the reflectivity spectra into absorbance units by an algorithm based on KKT and Fresnel's reflectivity equations. The absorbance conversion algorithm and the expression for d_p are not valid for nonhomogeneous samples, and therefore we will use them only for illustrative purposes.

Examples: Experimental Procedures and Calculations

To demonstrate the utility of the depth profiling method, we will analyze ATR FT-IR spectra collected at the film–air interface of a coalesced latex film (*28*). The sample film is about 100 μm thick and was prepared by drying a latex film on a poly(tetrafluoroethylene) surface. Dried latex film contains 4% w/w sodium dioctyl sulfosuccinate (SDOSS), used as a surfactant, and 96% w/w copolymer, which consists of 96% w/w ethyl acrylate (EA) and 4% w/w methacrylic acid (MAA) monomer units. The preparation procedures for the latex and film have been detailed elsewhere (*29*). It should be noted that EA/MAA latex films give spectra with a good signal-to-noise ratio because good contact with an ATR element can be maintained. This is because the glass transition temperature of such a copolymer is slightly below the temperature of an ATR experiment (about 5 °C).

ATR FT-IR spectra can be recorded on any FT-IR spectrometer equipped with a variable-angle multiple-reflection ATR accessory, which allows the external angles of incidence to be continuously varied from 30° to 60°. To achieve a wide range of depths of penetration, the following three parallelogram-shaped crystals can be used: $3 \times 20 \times 50$ mm^3 KRS-5 with a 45° bevel, $3 \times 20 \times 50$ mm^3 germanium with a 30° bevel, and $2 \times 20 \times 50$ mm^3 germanium with a 60° bevel. For many samples, using the KRS-5 crystal or the 30° beveled Ge crystal can result in saturated ATR spectra. To avoid such saturation, the area in contact with an ATR crystal should be reduced in the longitudinal direction of the crystal. In such a case, the sample length variation should be taken into account when the multiple-reflection ATR spectra are converted to reflectivity for one reflection. The incident light should be polarized perpendicular to the plane of incidence. It is also necessary to know the refractive index values in a sample transparent spectral region, which can be recorded on any refractometer—for example, using a sodium emission line in the ultraviolet and visible region.

Figure 7.2 shows ATR FT-IR spectra of a latex film and its components in the C–H stretching region. The spectra of the latex (traces B and C) contain distinct spectral features of the SDOSS surfactant and the EA/MAA copolymer (traces A and D, respectively). The spectra of SDOSS (trace A), EA/MAA copolymer (trace D), and latex film (trace C) were recorded using a KRS-5 crystal at an angle of 51.2°, while spectrum B (latex film) was recorded using the Ge crystal at 60°. The band height ratios of the bands at 2875 and 2985 cm^{-1} are listed in Table 7.1, and they can be used for the depth profile calculation for the nonhomogeneous surfaces. Because the optical theory applies at every wavenumber, whether the sample is absorbing or nonabsorbing in a particular spectral region, curve fitting is not required. In fact, curve fitting should not be used for ATR spectra because ATR intensities are not linear combinations of the component bands. Because neither of the two bands can be attributed to a single component, eqs 7.16 and 7.17 can be used to correlate the concentration of a given species with the optical constants of each layer. The optical constants for the pure SDOSS surfactant and the purified copolymer are required in the equations and are listed in Table 7.2. These values can be calculated from their corresponding ATR spectra using the Urban–Huang KKT algorithm described in Chapter 6, which, unlike the algorithms proposed in the past, can simultaneously handle weak and strong band intensities (*46*). The refractive index of the EA/MAA latex film at infinite wavenumber is estimated to be 1.4685.

The measured reflectivity data can be analyzed by the procedure diagrammed in Figure 7.3. The calculation begins by setting up a trial depth profile, followed by evaluation of the trial calculations. This is accompanied by comparing the calculated and measured ATR data. Figure 7.4 illustrates how the intensity ratios of the bands at 2875 and 2985 cm^{-1} (Y^{calc})

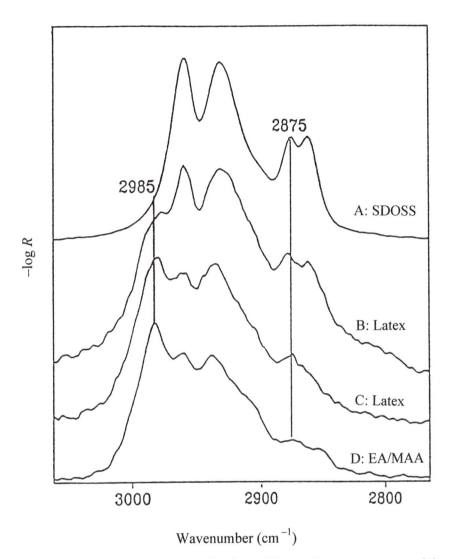

Figure 7.2. ATR FT-IR spectra of a latex film and its components: (A) SDOSS, KRS-5, 51.2°; (B) latex film, Ge, 60°; (C) latex film, KRS-5, 51.2°; (D) EA/MAA copolymer, KRS-5, 51.2°.

are calculated from a trial depth profile. The process is repeated in several embedded loops, and within each loop, a parameter used in setting up the trial depth profile is varied within a feasible range. When all the loops are completed, the depth profile that yields the smallest Q values based on a comparison of Y^{calc} and Y^{meas} in the diagram in Figure 7.3, and given by eq 7.19, is saved as the most probable depth profile. Note that the process

Table 7.1. Measured intensity ratios of the bands at 2875 and 2985 cm^{-1}

ATR crystal (n_1)	Angle of incidence (α_1)	Penetration depth[a] (d_p) at 2875 cm^{-1}	$\dfrac{-\ln R_{2875}{}^{b}}{-\ln R_{2985}}$
Germanium (4.001)	60.0	0.176	0.826
	56.9	0.184	0.820
	53.9	0.192	0.805
	37.2	0.288	0.690
	35.5	0.307	0.675
	33.7	0.332	0.619
	32.5	0.352	0.611
	31.2	0.377	0.590
KRS-5 (2.38)	51.2	0.488	0.453
	50.2	0.508	0.448
	49.2	0.531	0.463
	48.1	0.557	0.430
	47.1	0.589	0.413
	46.1	0.627	0.442
	45.0	0.673	0.407
	43.9	0.732	0.442
	42.9	0.809	0.430

[a] Not directly used in depth profiling calculation.
[b] R represents the reflectivity for one reflection.

Table 7.2. Optical constants of SDOSS and EA/MAA copolymer

	Wavenumber (cm^{-1})	EA/MAA copolymer	SDOSS
$k_2{}^\circ$	2875	0.013	0.063
	2985	0.057	0.027
$\Delta n_2{}^\circ$	2875	0.009	0.021
	2985	-0.035	-0.090

of setting up a trial depth profile consists of two steps. In the first step, the surface is divided into imaginary layers with the boundaries at a series of depths z_j at which the surfactant concentration ϕ_j is used to represent the concentration of the layer below. In the second step, assumptions are made about the relations between the concentrations of the various layers

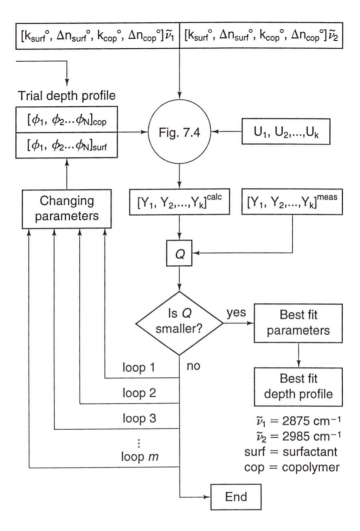

Figure 7.3. Flowchart for the fitting process for calculation of surfactant concentrations at a given depth in latex films.

to reduce the number of variables. Using this method, let us examine three situations, referred to as models A, B, and C.

Model A

The sample is divided into 40 equally spaced layers separated by 41 boundaries. The boundary spacing b and the concentrations at the first and last boundaries, ϕ_1 at $z_1 = 0$ and ϕ_{41} at $z_{41} = 40b$, are varied in three embed-

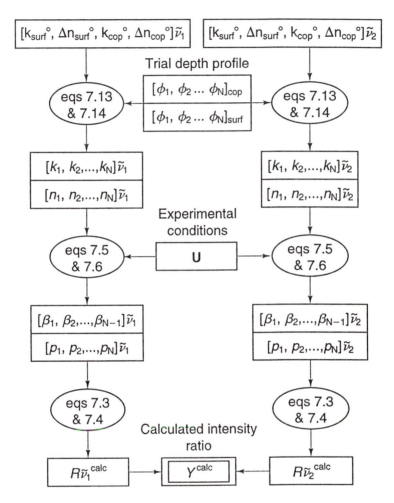

Figure 7.4. Flowchart for the calculation of reflectivity data from a trial depth profile.

ded loops, and the concentrations at the remaining boundaries are linearly interpolated.

In this approach, the dividing boundaries are located at $z_j = (j-1)b$, $j = 1, 2, ..., 41$, where the spacing b is a fitting variable. The concentrations ϕ_j at the 2nd through the 40th boundaries are linearly interpolated from ϕ_1 at $z_1 = 0$ to ϕ_{41} at $z_{41} = 40b$ according to the following equation:

$$\phi_j = \phi_1 + (\phi_{41} - \phi_1)\frac{j-1}{40} \qquad j = 2, 3, ..., 40 \qquad (7.20)$$

The parameters ϕ_1, ϕ_{41}, and b are varied in three loops. In loop 1, ϕ_1 is varied according to the following recurrence formula:

$$\phi_1^j = \phi_1^{j-1} \frac{\sum\limits_{l=1}^{k} Y_l^{meas}}{\sum\limits_{l=1}^{k} Y_l^{calc}} \tag{7.21}$$

where the superscript of ϕ_1 denotes the stage of iteration, Y_l^{calc} is the ATR intensity ratio of the bands at 2875 and 2985 cm^{-1} calculated by eq 7.15, and Y_l^{meas} is the measured intensity ratio of the baseline-corrected bands. This loop will continue until ϕ_1 varies by less than 0.001. The iteration based on eq 7.21 is expected to converge because the measured ATR band intensities increase with the increasing surfactant concentrations at the surface. In loop 2, ϕ_{41} varies from 0 to 1 with an increment of 0.001. Similarly, in loop 3, the boundary spacing h varies from 0.00025 to 0.025 μm in increments of 0.00025. This corresponds to a range of the depths of the last boundary from 0.01 to 1 μm.

Model B

In this case, the sample is divided into six primary layers separated by boundaries with variable spacings. The layers between the primary boundaries will be divided into five thinner sublayers of equal thickness. Volume fractions at the primary boundaries are varied in six embedded loops, and the concentrations at the sublayer boundaries are linearly interpolated.

The dividing boundaries are located at the depths given by

$$
\begin{aligned}
z_j &= (j-1)h_1/5 & j &= 1, 2, ..., 6 \\
z_j &= z_6 + (j-6)h_2/5 & j &= 7, 8, ..., 11 \\
z_j &= z_{11} + (j-11)h_3/5 & j &= 12, 13, ..., 16 \\
z_j &= z_{16} + (j-16)h_4/5 & j &= 17, 18, ..., 21 \\
z_j &= z_{21} + (j-21)h_5/5 & j &= 22, 23, ..., 26
\end{aligned}
\tag{7.22}
$$

and

$$h_s = h_1 e^{(s-1)d} \qquad s = 1, 2, ..., 5 \tag{7.23}$$

where $h_1 = 0.05$ and $d = 0.227$. According to this method of division, the boundary spacing varies from 0.01 μm near the surface to 0.025 μm near the last boundary, located at $z_{40} = 0.51$ μm. The depth of the last boundary is about 3.5 times the optimum depth of the last boundary (0.14 μm) calculated in model A. The reason for choosing a variable boundary spacing is that the reflectivity data are more sensitive to concentration variations that are nearer the sample surface. The concentrations at various boundaries are linearly interpolated from the volume fractions at the primary boundaries located at z_1, z_6, z_{11}, z_{16}, z_{21}, and z_{26} according to the following relations:

$$\begin{aligned}
\phi_j &= \phi_1 + (\phi_6 - \phi_1)(j-1)/5 & j &= 2, 3, \ldots, 6 \\
\phi_j &= \phi_6 + (\phi_{11} - \phi_6)(j-6)/5 & j &= 7, 8, \ldots, 11 \\
\phi_j &= \phi_{11} + (\phi_{16} - \phi_{11})(j-11)/5 & j &= 12, 13, \ldots, 16 \qquad (7.24) \\
\phi_j &= \phi_{16} + (\phi_{21} - \phi_{16})(j-16)/5 & j &= 17, 18, \ldots, 21 \\
\phi_j &= \phi_{21} + (\phi_{26} - \phi_{21})(j-21)/5 & j &= 22, 23, \ldots, 26
\end{aligned}$$

To further reduce the calculation time, we assume that the surfactant concentration decreases monotonically with increasing depth and therefore $\phi_1 > \phi_6 > \phi_{11} > \phi_{16} > \phi_{21} > \phi_{26}$. With this assumption, the concentrations at the primary boundaries will be varied according to the following scheme:

$$\phi_j = \phi_{j-5} - \delta_j \qquad j = 6, 11, 16, 21, 26 \qquad (7.25)$$

where δ_j represents the variation of the primary boundary concentration relative to the concentration at the previous primary boundary. The parameters ϕ_1 and δ_j are varied in six embedded loops. In loop 1, ϕ_1 is varied by the recurrence formula in eq 7.21. This loop is continued until ϕ_1 varies by less than 0.001. In each of the remaining five loops, the value for δ_j is varied at an increment of 0.05 in the following range:

$$0 \le \delta_j \le 1 - \sum_{s=\mathrm{int}(j/5)+1}^{5} \delta_{5s+1} \qquad j = 6, 11, 16, 21 \qquad (7.26)$$

The upper limits are set by the requirement that all concentrations be non-negative.

Model C

The sample is divided into 40 layers, as in model A, except that the boundary spacings are fixed at 0.01 μm, estimated from the results of model A. The volume fractions at the first and the last boundaries, ϕ_1 and ϕ_{41}, are treated as the independent variables for fitting and the concentrations at the remaining boundaries are exponentially interpolated using the following equation:

$$\phi_j = (\phi_1 - \phi_{41})e^{-z_j/d_\phi} + \phi_{41} \qquad j = 2, \ldots, 40 \qquad (7.27)$$

where d_ϕ is another variable for the fitting process.

In this model, the 41 boundaries are located at the depths z_j given by eq 7.20 with $b = 0.01$ μm, which is about three times the optimum value obtained in model A. The concentrations at the first and last boundaries, ϕ_1 and ϕ_{41}, as well as d_ϕ in eq 7.27, are varied in three embedded loops. In loop 1, ϕ_1 is varied by the recurrence formula in eq 7.21. This loop is continued until ϕ_1 varies by less than 0.001. In loop 2, ϕ_{41} is varied from 0 to 1 in increments of 0.001. In loop 3, d_ϕ is varied from 0.01 to 2 μm in increments of 0.01 μm.

Accuracy of Calculations and Summary

In the three models, all boundaries will be located within the maximum. The experimentally established penetration depth varies from sample to sample (0.8 μm for EA/MAA latex at 1000 cm^{-1}) because the surfactant concentrations beyond the maximum penetration depth cannot be calculated from the measured reflectivity data. Furthermore, the copolymer volume fractions are given by $\phi_j^{cop} = 1 - \phi_j$.

The calculated surfactant depth profiles for all three methods are shown in Figure 7.5, where the error bars (±0.05) for model B represent the increment by which volume fractions of the leading sublayers are individually varied between 0 and 1. Because models A and C contain variable parameters that are not volume fractions, no appropriate error bars can be identified on the curves for these models. Instead, the variable ranges and increments for these parameters are listed in Table 7.3. As shown in Figure 7.5, the results obtained from the three models are similar.

It is not surprising that in almost every fitting procedure—and the approach presented is no exception—as the number of fitting parameters increases, the fit improves. Therefore, it is important to evaluate the sensitivity of the developed procedure to the number of fitting parameters. In our case, Figure 7.6 illustrates the calculated absorbance ratios plotted as a function of the penetration depth at 2875 cm^{-1}, along with the measured intensity ratios from models A, B, and C. Since the depth profiles calculated using all models are similar, their calculated reflectivity values are also similar and fit the measured data fairly well. Although the assumptions used to calculate the unknown depth profiles are valid for the surfactant depth profiles in Figure 7.6, it should be realized that the three proposed models explore a new method for calculating surface depth profiles using optical theory and should serve as examples to be tested for other systems. The flexibility of the proposed approach comes from the fact that if no satisfactory fit is found for a given system, new models can be devised, and the same general method based on trying and testing the data using optical theory will remain valid. As long as the number of unknown parameters is smaller than the number of the independent equations, and as long as the

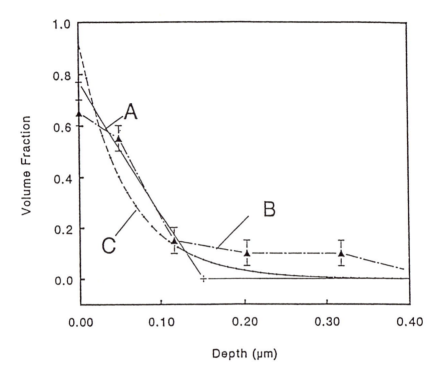

Figure 7.5. Calculated surfactant depth profiles in EA/MAA latex coatings: (A) linear interpolation with two primary layers; (B) linear interpolation with six primary layers; (C) exponential model.

Table 7.3. Variable ranges and increments for models A and C

Parameter	Models	From	To	Increment
ϕ_1	A & C	0	1	0.001
ϕ_{41}	A & C	0	1	0.001
b (μm)	A	0.00025	0.025	0.00025
d_ϕ (μm)	C	0.01	2	0.01

number of equations is no greater than the number of experimental data points, the optimized parameters are valid.

Reflection theory for stepwise stratified media can be applied to the reflectivity data obtained in the ATR depth profiling experiments. Depth profiling ATR measurements can be used to estimate the concentration changes at various depths. The experimental parameters, such as the refractive index of ATR crystals and the angles of incidence, are used

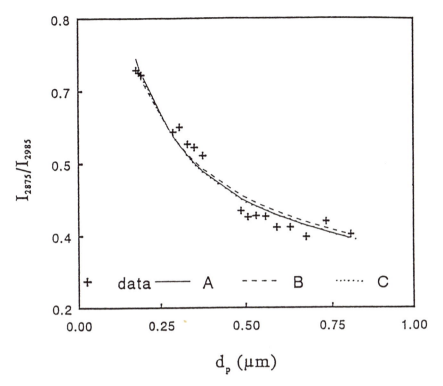

Figure 7.6. Measured and calculated intensity ratios of the bands at 2875 and 2985 cm^{-1} as a function of penetration depth: (A) linear interpolation with two primary layers; (B) linear interpolation with six primary layers; (C) exponential model; scattered points are measured data.

directly, without converting them to penetration depths. As a result, the expressions for penetration depth dependence for weak bands can be avoided. Furthermore, this theory also takes into account the effect of refractive index dispersion, which is otherwise intractable for nonhomogeneous samples. The unknown depth profile can be calculated by numerical methods, including linear interpolation and the use of an assumed function with variable parameters as a trial depth profile.

Thin Films and Coatings

When thin films or coatings are homogeneous and have a thickness comparable to or smaller than the penetration depth, the reflectance can be also calculated from the multiphase reflection theory. If the sample thickness is much smaller than the penetration depth, we can assume that the electric field is constant within the film thickness and equal to the value calculated as if the sample were absent. Under such circumstances, the ATR intensity is proportional to the linear absorptivity and the film thickness:

$$-\ln R_\perp = \beta d\left(\frac{4n_1 n_2 \cos\alpha_1}{n_1^2 - n_b^2}\right) \qquad (7.28)$$

and

$$-\ln R_\parallel = \beta d\left(\frac{4n_1 n_2 \cos\alpha_1 [(n_2^4 + n_{2,b}^4)n_1^2 \sin^2\alpha_1 - n_{2,b}^2 n_2^4]}{(n_1^2 - n_{2,b}^2)n^4 [(n_1^2 + n_{2,b}^2)\sin^2\alpha_1 - n_{2,b}^2]}\right) \qquad (7.29)$$

where the subscript b denotes the backing, and $\perp$ and $\parallel$ denote s- and p-polarization, respectively.

Examination of the expression for $-\ln R_\perp$ indicates that the ATR intensity increases with increasing refractive indices of the sample and its backing and with decreasing angle of incidence.

In an effort to establish the ranges of validity for the proportionality between ATR band intensity and film thickness or linear absorptivity, two sets of simulation curves using the matrix theory for s-polarized light reflection were constructed. In the first set (Figures 7.7 and 7.8), the ATR band intensity divided by the linear absorptivity is plotted as a function of film thickness; in the second set (Figures 7.9 and 7.10), the ATR band intensity divided by the film thickness is plotted as a function of linear absorptivity. Figures 7.7 and 7.9 were constructed for the KRS-5–film–air configuration, whereas for Figures 7.8 and 7.10, the air backing was replaced with transparent material whose refractive index matches that of the sample in a nonabsorbing region. Note that in Figures 7.7 through 7.10, the ATR band intensity is defined as the maximum of $-\ln R(\tilde{v})$ over a wavenumber range from 1730 to 1670 cm^{-1}, where the hypothetical sample exhibits an absorption maximum at 1700 cm^{-1}. This issue is particularly important because the frequency shift of the ATR spectrum relative to the absorbance spectrum means that the ATR intensity is not measured at 1700 cm^{-1}. This definition of ATR band intensity is relevant to practical measurement and therefore applies to all the simulation results provided later.

As indicated in Figures 7.7 and 7.8, the deviation from proportionality between ATR band intensity and film thickness begins at very small film thicknesses. For this example, it extends no further than 0.2 μm. In contrast, Figures 7.9 and 7.10 illustrate that the proportionality between ATR band intensity and linear absorptivity extends up to fairly high intensities.

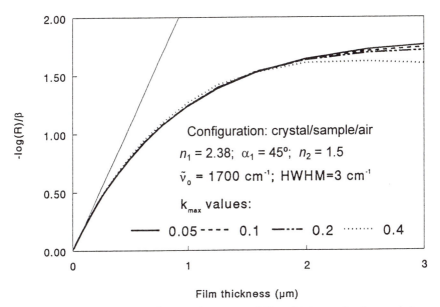

Figure 7.7. Plots of $-\ln R/\beta$ *as a function of film thickness for* $k_{2,max}$ *values ranging from 0.05 to 0.4. Configuration: KRS-5 crystal/sample (* $n_{2,\infty}$ = 1.5)/ *air (* $n_{2,b}$ = 1); angle of incidence α_1 = 45°; $\tilde{\nu}$ = 1700 cm^{-1}; HWHM = 3 cm^{-1}.

To improve the accuracy of eqs 7.28 and 7.29, and to avoid the complexity of the matrix representations, a new expression for the ATR band intensity for thin films was derived.

Since the ATR band intensity is zero at zero thickness and approaches the ATR intensity of the bulk as film thickness goes to infinity, the following empirical expression for the ATR intensity of thin film samples may be appropriate:

$$-\ln R_i = A_i(1 - e^{-B_i d}) \tag{7.30}$$

where $i = \perp$ or $\parallel$ (s- or p-polarization), d is the film thickness, A_i is the ATR intensity of the bulk sample given by Fresnel's equations, and B_i is a parameter to be determined. Setting the initial slope of $-\ln R$ versus film thickness to the value given by eqs 7.28 and 7.29, we obtain

$$B_\perp = \frac{\beta}{A_\perp}\left(\frac{4n_1 n_2 \cos\alpha_1}{n_1^2 - n_{2,b}^2}\right) \tag{7.31}$$

and

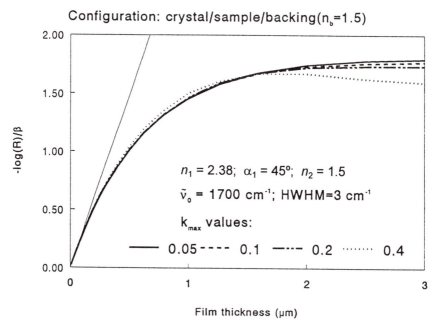

Figure 7.8. Plots of $-\ln R/\beta$ as a function of film thickness for $k_{2,max}$ values ranging from 0.05 to 0.4. Configuration: KRS-5 crystal/sample ($n_{2,\infty}$ = 1.5)/ transparent backing ($n_{2,b}$ = 1.5); angle of incidence α_1 = 45°, $\tilde{v}$ = 1700 cm^{-1}; HWHM = 3 cm^{-1}.

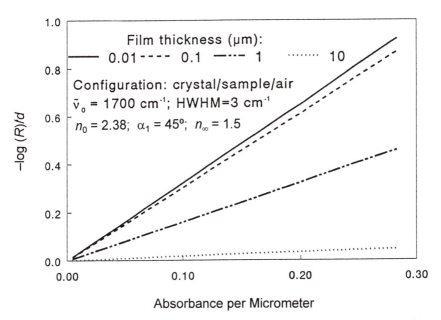

Figure 7.9. Plots of −ln R/d as a function of linear absorptivity for film thicknesses ranging from 0.01 to 10 μm. Configuration: KRS-5 crystal/sample ($n_{2,\infty}$ = 1.5)/air ($n_{2,b}$ = 1); angle of incidence α_1 = 45°; $\tilde{v}$ = 1700 cm^{-1}; HWHM = 3 cm^{-1}.

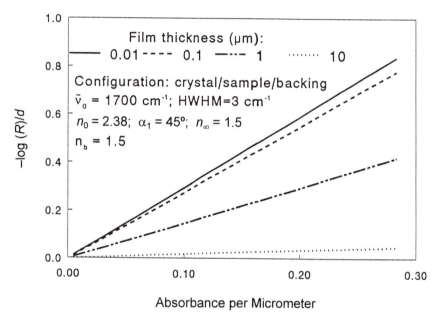

Figure 7.10. Plot of −ln R/d as a function of absorbance for film thicknesses ranging from 0.01 to 10 µm. Configuration: KRS-5 crystal/sample ($n_{2,\infty}$ = 1.5)/transparent backing ($n_{2,b}$ = 1.5); angle of incidence α_1 = 45°, $\tilde{v}$ = 1700 cm^{-1}; HWHM = 3 cm^{-1}.

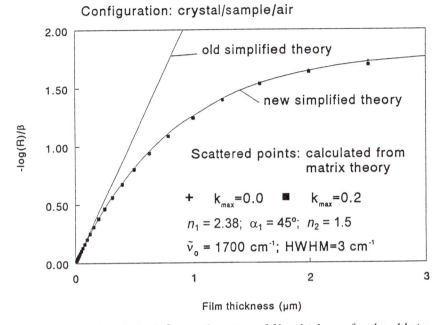

Figure 7.11. Plot of $-\ln R/\beta$ as a function of film thickness for the old simplified model and the new theory.

$$B_{\parallel} = \frac{\beta}{A_{\parallel}}\left(\frac{4 n_1 n_2 \cos\alpha_1 [(n_2^4 + n_{2,b}^4) n_1^2 \sin^2\alpha_1 - n_{2,b}^2 n_2^4]}{(n_1^2 - n_{2,b}^2) n^4 [(n_1^2 + n_{2,b}^2)\sin^2\alpha_1 - n_{2,b}^2]}\right) \qquad (7.32)$$

To summarize, according to the literature, the relationship between ATR band intensity and film thickness is linear. This theory indeed holds true for very thick films, but not for thin films. Figure 7.11 is a plot of $(-\ln R)/\beta$ versus film thickness comparing the simplified approach used in the past and the approach presented in this chapter.

References

1. Blackwell, C. S.; Degen, P. J.; Osterholtz, F. D. *Appl. Spectrosc.* **1978**, *32*, 480.

2. Carlsson, D. J.; Wiles, D. M. *Macromolecules* **1972**, *4*, 173.

3. Webb, J. R. *J. Polym. Sci. Polym. Chem. Ed.* **1972**, *10*, 2335.

4. Hirayama, T.; Urban, M. W. *Prog. Org. Coat.* **1992**, *21*, 81.

5. Harrick, N. J. *J. Phys. Chem.* **1960**, *64*, 1110.

6. Urban, M. W.; Huang, J. B. *Appl. Spectrosc.* **1993**, *47*, 973.

7. Harrick, N. J. *Internal Reflection Spectroscopy;* Interscience Publishers: New York, 1967.

8. Mirabella, F. M. *Spectroscopy* **1990**, *5*, 21.

9. Koenig, J. L.; Esposito, D. L.; Antoon, M. K. *Appl. Spectrosc.* **1977**, *31*, 292.

10. Mirabella, F. M. *J. Polym. Sci. Polym. Phys. Ed.* **1985**, *30*, 861.

11. Mirabella, F. M. *Appl. Spectrosc. Rev.* **1985**, *21*, 45.

12. Crawford, B., Jr.; Goplen, T. G.; Swanson, D. In *Advances in Infrared and Raman Spectroscopy;* Clark, R. J. H.; Hester, R. E., Eds.; Heyden: London, 1980; Vol. 4, Chapter 2.

13. Graf, R. T.; Koenig, J. L.; Ishida, H. In *Fourier Transmission Infrared Characterization of Polymers;* Ishida, H., Ed.; Plenum: New York, 1987.

14. Tompkins, H. G. *Appl. Spectrosc.* **1974**, *28*, 335.

15. Harrick, N. J. *J. Opt. Soc. Am.* **1965**, *55*, 851.

16. Carlsson, D. J.; Wiles, D. M. *Can. J. Chem.* **1970**, *48*, 2397.

17. Harrick, N. J.; Du Pre, F. K. *Appl. Opt.* **1966**, *5*, 1739.

18. Harada, M.; Kitamori, T.; Teramae, N.; Hashimoto, K.; Oda, S.; Sawada, T. *Appl. Spectrosc.* **1992**, *46*, 529.

19. Hirschfeld, T. *Appl. Spectrosc.* **1977**, *31*, 289.

20. Bellman, R.; Kalaba, R. E.; Lockett, J. A. *Numerical Inversion of the Laplace Transform;* Elsevier: New York, 1966.

21. Krylov, V. I.; Skoblya, N. S. *Handbook of Numerical Inversion of Laplace Transforms* (English transl.); Israel Program for Scientific Translations: Jerusalem, 1969.

22. Fina, L. J.; Chen, G. C. *Vibr. Spectrosc.* **1991**, *1*, 353.

23. Hansen, W. *J. Opt. Soc. Am.* **1968**, *58*, 380.

24. Stuchebryukov, S. D. *Surf. Interface Anal.* **1984**, *6*(1), 29.

25. Fina, L. *Appl. Spectrosc. Rev.* **1994**, *29*(3–4), 309.

26. Heavens, O. S. *Thin Film Physics;* Methuen: London, 1970.

27. Born, M.; Wolf, E. *Principles of Optics,* 5th ed.; Pergamon: Oxford, 1975.

28. Urban, M. W.; Evanson, K. W. *Polym. Comm.* **1990**, *31*, 279.

29. Evanson, K. W.; Urban, M. W. *J. Appl. Polym. Sci.* **1991**, *42*, 2287.

30. Evanson, K. W.; Thorstenson, T. A.; Urban, M. W. *J. Appl. Polym. Sci.* **1991**, *42*, 2297.

31. Evanson, K. W.; Urban, M. W. *J. Appl. Polym. Sci.* **1991**, *42*, 2309.

32. Evanson, K. W.; Urban, M. W. In *Surface Phenomena and Fine Particles in Water-Based Coatings and Printing Technology;* Sharma, M. K.; Micele, F. J., Eds.; Plenum: New York, 1991.

33. Thorstenson, T. A.; Urban, M. W. *J. Appl. Polym. Sci.* **1993**, *47*, 1381.

34. Thorstenson, T. A.; Urban, M. W. *J. Appl. Polym. Sci.* **1993**, *47*, 1387.

35. Thorstenson, T. A.; Tebelius, L. K.; Urban, M. W. *J. Appl. Polym. Sci.* **1993**, *49*, 103.

36. Thorstenson, T. A.; Tebelius, L. K.; Urban, M. W. *J. Appl. Polym. Sci.* **1993**, *50*, 1207.

37. Kungel, J. P.; Urban, M. W. *J. Appl. Polym. Sci.* **1993**, *50*, 1217.

38. Tebelius, L. K.; Urban, M. W. *J. Appl. Polym. Sci.* **1995**, *56*, 387.

39. Niu, B.; Urban, M. W. *J. Appl. Polym. Sci.* **1995**, *56*, 377.

40. Folkes, G. R. *Introduction to Modern Optics;* Holt, Rinehart, and Winston: New York, 1975.

41. Frohlish, H. *Theory of Dielectrics;* Clarendon: Oxford, 1949.

42. Maeda, S.; Schatz, P. N. *J. Chem. Phys.* **1962**, *34*, 571.

43. Hansen, W. N.; Abdou, W. A. *J. Opt. Soc. Am.* **1977**, *67*, 1537.

44. Goplen, T. G.; Cameron, D. G.; Jones, R. N. *Appl. Spectrosc.* **1980**, *34*, 657.

45. Graf, R. T.; Koenig, J. L.; Ishida, H. *Appl. Spectrosc.* **1985**, *39*, 405.

46. Huang, J. B.; Urban, M. W. *Appl. Spectrosc.* **1992**, *46*, 1666.

Section III

Selected Applications

8

ATR of Surfaces and Interfaces

Analysis of Polymer Surfaces

ATR spectroscopy is a useful method for analyzing surfaces of elastomeric materials with low glass transition temperatures. Such materials are soft, and good contact between the sample surface and the crystal is easily maintained. Therefore, highly reproducible results with a high signal-to-noise ratio are expected. In contrast, if a rough and rigid sample is analyzed, the optical contact is usually poor, so the signal-to-noise ratio is smaller. The best surface is the surface of a fluid, and perhaps this fact accounts for the use of in situ ATR to monitor the kinetics of polymerization.

Polyethylene

Although numerous papers are published every year dealing with IR spectroscopy of polyethylene, only a few address the structures that develop on polyethylene surfaces. In fact, the primary focus of many studies has been on spectroscopic determination of crystallinity and defect structures in polyethylene. The crystalline content affects and dictates many macroscopic properties, so its determination is important. In the initial spectroscopic studies, crystallinity was determined by ratioing the intensity of the 1303 cm^{-1} band due to the amorphous phase in the solid and molten states (1). Because this approach uses one band in two spectra (solid and molten), it suffers from uncertainties attributable to the differences between absorption coefficients in the solid and molten states. Furthermore, uncontrollable changes of the film thickness on melting may be a source of error. Later on, measurements involving one band in one spectrum were developed for the amorphous band at 1303 cm^{-1}, for which a universal calibration constant was established (2). In this case, crystallinity calculations also require knowing the thickness and film density. This approach was extended to other bands, including the band at 730 cm^{-1} due to CH$_2$ rocking normal vibrations of the crystalline phase (3). The 730 cm^{-1} band, how-

3348–9/96/0135/$22.00/0/© 1996 American Chemical Society

ever, did not perform as well because of its sensitivity to the polymer chain orientation (3). Although in all cases baseline correction and curve fitting procedures are required (4), it is generally accepted that a more convenient and reliable method involves the use of two bands due to the crystalline phase at 1894 cm^{-1} and the band at 1303 cm^{-1} due to the amorphous phase (5).

While measurements of the properties of bulk polyethylene have been and continue to be of importance, the surface content is also of interest because it may affect adhesion, durability, and other macroscopic functions. For that reason, topological inhomogeneity and distribution of the crystalline and amorphous phases on polyethylene surfaces have been the subject of studies (6, 7) in which ATR FT-IR spectroscopy was the technique of choice.

Like the IR transmission spectrum, the ATR spectrum of polyethylene contains two bands due to pure crystalline (1894 cm^{-1}) and pure amorphous (1303 cm^{-1}) normal modes, but unfortunately these bands are too weak to be used for ATR quantitative analysis. Therefore, more intense bands, such as CH$_2$ bending (1473 and 1464 cm^{-1}) and rocking (730 and 720 cm^{-1}) normal vibrations were employed (6). The bands at 1474 and 730 cm^{-1} are due to the crystalline phase only, whereas the bands at 1464 and 720 cm^{-1} represent contributions from both crystalline and amorphous phases. These bands can be used to determine the amorphous content x by means of the following empirical relationship (6):

$$
\begin{aligned}
x &= \frac{I_b - I_a/1.233}{I_a + I_b} \times 100\% \\
&= \frac{1 - (I_a/I_b)/1.233}{1 + I_a/I_b} \times 100\%
\end{aligned}
\tag{8.1}
$$

where I_a and I_b are the intensities of the 730 and 720 cm^{-1} or 1474 and 1464 cm^{-1} bands, depending on which pair is chosen for analysis. The constant 1.233 in eq 8.1 is the intensity ratios of these bands in the spectrum of pure crystalline polyethylene and was derived by applying factor group splitting to a single polyethylene crystal (7). The key spectral features used in this method are the experimentally determined intensity ratio I_a/I_b, obtained from the spectrum of the semicrystalline sample, and the theoretical values of the ratios for pure crystalline polyethylene. This approach was applied to commercial polyethylene films (6), and it was concluded that the crystallinity of polyethylene was considerably higher near the surface. However, two issues that may significantly affect the intensity ratio are (1) distortion of intense IR bands in ATR spectra caused by optical dispersions and (2) orientation of the crystallites on the polyethylene surface (8).

Let us further analyze an ATR spectrum of polyethylene shown in Figure 8.1, trace A. The spectrum was recorded at a 45° incident angle with

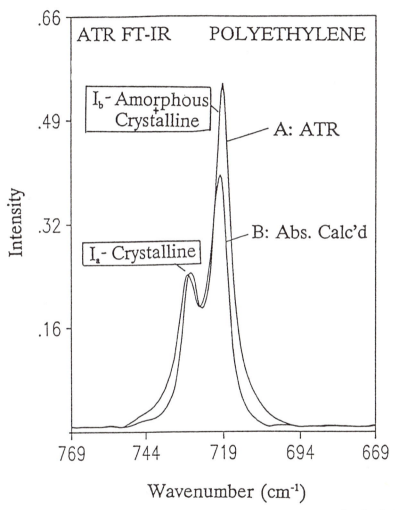

Figure 8.1. (A) raw and (B) corrected ATR FT-IR spectrum of polyethylene recorded at a 45° angle of incidence with the sample machine direction perpendicular to the electric vector of the s-polarized light. (Reproduced with permission from reference 8. Copyright 1992 Butterworth–Heinemann, Ltd.)

the sample machine direction perpendicular to the electric vector of the s-polarized light. Trace B of Figure 8.1 is the absorbance spectrum calculated from trace A by the Urban–Huang algorithm (Chapter 5). It is apparent that the intensity ratio of the 730 and 720 cm^{-1} bands in trace A is much greater than that obtained after the transformation represented by trace B.

According to eq 7.1, by ratioing the ATR spectrum to the absorbance spectrum, we obtain the effective thickness dependence. Therefore, the difference between the ATR spectrum and the true absorbance spectrum can be predicted from eqs 7.2 and 7.3 by using a nonconstant refractive index. Trace B of Figure 8.2 is the refractive index spectrum of polyethylene, calculated from trace A of Figure 8.2, which is the same spectrum as that shown in trace A of Figure 8.1. The effective thickness spectra obtained by using eq 6.2 for various angles of incidence and the same refractive index spectrum (trace B of Figure 8.2) are shown in Figure 8.3. The spectra illustrate that, for all angles of incidence, the effective thickness of the 730 cm^{-1} band is smaller than that at 720 cm^{-1}, and unequal effective thicknesses at these two bands are responsible for the different intensity ratios measured from traces A and B of Figure 8.1. The spectral distortion of trace A relative to trace B of Figure 8.1 is an example of perturbation between the two neighboring absorption bands. The first general discussion of this subject was by Hawranek and Jones (9), who showed that the extent of perturbation increases with the maximum absorbance, the band overlap of the adjacent IR bands, and the average refractive index of the sample. These issues were discussed in Chapter 4.

Let us now examine how optical dispersions may affect distortion of ATR strong bands that, in turn, will influence the surface crystallinity measurements in polyethylene. To see the difference in the amorphous phase content, we will calculate this quantity using raw ATR spectra intensities following procedures employed by Abbate et al. (7) and using the algorithm in Figure 6.5 that employs double KKT and Fresnel's equation to treat ATR intensities. In both cases, eq 8.1 will be used. If the 730 and 720 cm^{-1} bands are separated by using Lorentzian curve fitting and the calculated amorphous content is plotted as a function of the penetration depth, Figure 8.4 results. Curves A and B were obtained from the raw ATR spectra recorded in the sample machine direction aligned perpendicular and parallel to the electric vector of the s-polarized light, respectively. Curves A' and B' are the counterparts of A and B but were obtained by applying KKT and Fresnel's equation to the raw ATR spectra. As might be expected, significant intensity ratio differences between the raw and corrected ATR spectra lead to differences in the amorphous contents calculated from them. These results indicate again that a proper combination of KKT and Fresnel's relationship leads to results substantially different from those obtained by direct intensity measurements. Although differences in surface crystallinity content may be responsible for these discrepancies, the use of

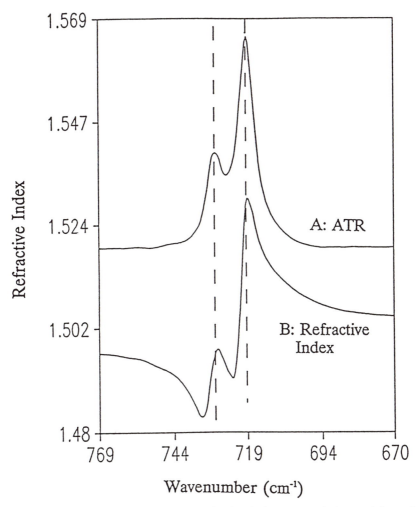

Figure 8.2. (A) ATR FT-IR spectrum of polyethylene recorded at a 45° angle of incidence and (B) calculated real part of the refractive index spectrum. (Reproduced with permission from reference 8. Copyright 1992 Butterworth–Heinemann, Ltd.)

KRS-5 crystal in contact with polyethylene cannot give a 15 μm penetration depth, as claimed by Dothee et al. (6), unless the measurements are taken at an angle of incidence near the critical angle. Under such circumstances, the bands will experience even more intensity distortions due to optical effects and therefore should not be used for quantification of the surface crystallinity content in polyethylene. This issue has been addressed in various studies (10–12), and ignorance of the optical effects for strong overlapping bands often leads to misleading conclusions. A perfect example is a

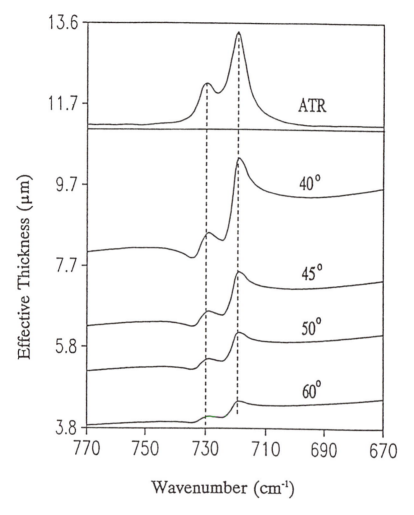

Figure 8.3. Effective thickness as a function of wavenumber for various angles of incidence calculated from Figure 8.2B. (Reproduced with permission from reference 8. Copyright 1992 Butterworth–Heinemann, Ltd.)

direct use of the intensity ratios I_a/I_b of the CH_2 bending or rocking normal vibrations for surface crystallinity measurements in polyethylene.

Polypropylene

Surface orientation of polypropylene subjected to uniaxial and biaxial elongations was extensively investigated in a series of pioneering studies by

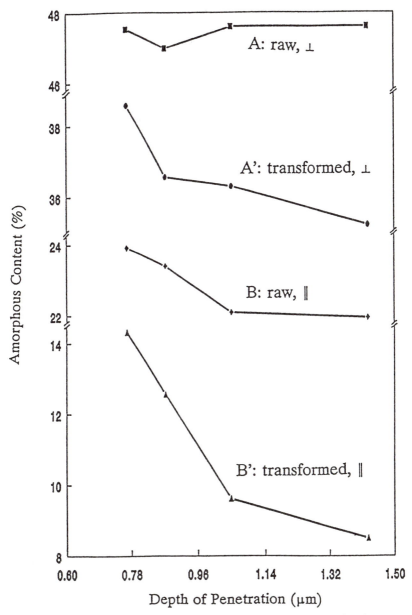

Figure 8.4. Amorphous content (%) as a function of the depth of penetration calculated from ATR spectra: (A) raw perpendicular polarization, (A') corrected perpendicular polarization, (B) raw parallel polarization, and (B') corrected parallel polarization. (Reproduced with permission from reference 8. Copyright 1992 Butterworth–Heinemann, Ltd.)

Mirabella (*13, 14*). Apparently the 841 and 809 cm^{-1} bands can be used to identify the surface orientation of the helix axis of the polypropylene. The intensity ratio of these bands increases when the films are oriented in the elongation direction and decreases when the films are oriented in the transverse direction or normal to the film thickness. These studies stimulated many later studies involving stretching polymers and monitoring their orientation. Ethylene–propylene blend studies using near-IR ATR FT-IR spectroscopy showed promising results, but the question of quantitative analysis is yet to be resolved (*15*). Walls (*16*) critically examined the use of polarized ATR spectroscopy in an effort to examine surface composition and orientational effects in uniaxially drawn poly(ethylene terephthalate) (PET) films. This study showed that it is of primary importance to establish a reference band for monitoring phase changes in polymers. In the case of PET, such a band is at 1410 cm^{-1}. Previously suggested reference bands at 790 and 1510 cm^{-1} appear to be influenced by both light polarization and orientation.

Poly(vinylidene fluoride)

The reactions of poly(vinylidene fluoride) (PVDF) in aqueous sodium hydroxide in the presence of tetrabutylammonium hydrogen sulfate (a phase transfer catalyst) result in surface dehydrofluorination (*17–19*). Although such reactions should result in a fully conjugated polymer backbone, head-to-head and tail-to-tail defects in the PVDF backbone result in noncontinuous conjugations formed of 30 to 50 carbon–carbon bonds, thus diminishing electrical conductivity (*20, 21*). Two procedures allowing grafting of an axially functionalized silicone phthalocyanine dichloride across a dihydrofluorinated PVDF surface were also examined by ATR (*22*).

Poly(vinyl chloride)

It appears that there are significant differences between polymers with the same repeating unit. For a polymer chemist this is no surprise because it is well known that even small changes in molecular weight or functionality may result in drastic changes in properties. However, it is quite surprising that when poly(vinyl chloride) (PVC) is dissolved in tetrahydrofuran solvent and cast on a substrate, significantly different chemistries are detected for PVC from different suppliers. Figure 8.5 illustrates ATR FT-IR spectra of PVC obtained from two suppliers. Although one could argue about various spectral features, there is no reason for trace A to have a carbonyl stretching band at 1734 cm^{-1}, unless a substantial amount of plasticizer was added. On the other hand, Figure 8.6 illustrates how thermal history may

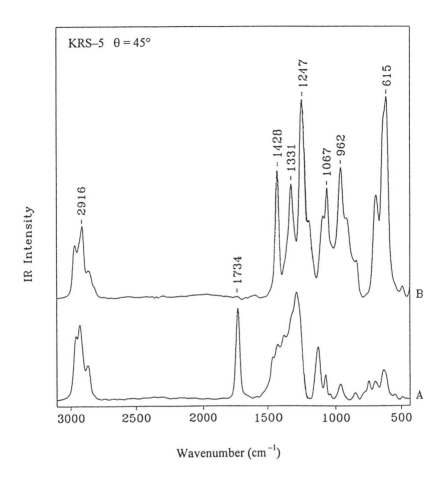

Figure 8.5. ATR FT-IR spectrum of solvent-cast (tetrahydrofuran) poly(vinyl chloride) from two supplies: (A) Alpha Chemicals & Plastics Co. and (B) Aldrich Co.

affect spectral, and therefore structural, features of PVC. Trace A shows the ATR FT-IR spectrum of hot-pressed PVC. Trace B was recorded from the same PVC specimen except that film was obtained by solvent casting.

The surface modifications of PVC appear to depend strongly on its thermal history. For example, Figure 8.7 shows a series of ATR spectra of imidazole and PVC reacted in the presence of microwave plasma. The two new bands at 1658 and 1587 cm^{-1} can be attributed to the I band of the imidazole ring and to C=N stretching. The resulting surface reactions are shown in Figure 8.8 (*23*).

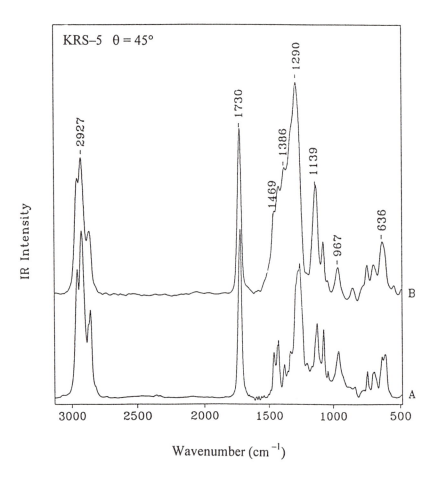

Figure 8.6. ATR FT-IR spectrum of poly(vinyl chloride) from Alpha Chemicals & Plastics Co.: (A) hot-pressed and (B) solvent-cast from tetrahydrofuran solution.

Urethanes

Urethane network formation can be monitored by following isocyanate consumption. In this case, the isocyanate band at 2270 cm^{-1} may serve as a probe of the reaction. Figure 8.9 illustrates two ATR FT-IR spectra recorded 24 hours (trace A) and 80 days (trace B) after the reaction of isocyanate with polyol at 40% relative humidity. Quantitative analysis of the isocyanate reactions is possible with a calibration curve (Figure 8.10), which correlates ATR corrected intensity with the actual concentration of the isocyanate groups. Figure 8.11 demonstrates how ATR concentration can

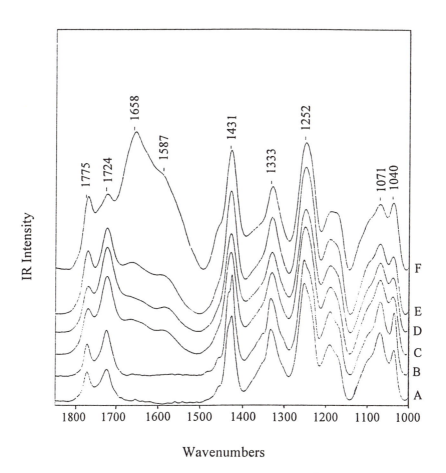

Figure 8.7. ATR FT-IR spectra in the 1850–1000 cm^{-1} region of imidazole reacted to solvent-cast PVC for 5 s in the presence of oxygen microwave plasma at various pressures using a closed reactor: (A) unreacted solvent-cast PVC, (B) oxygen microwave plasma at 106.4 Pa, (C) imidazole reacted to PVC at 106.4 Pa, (D) imidazole reacted to PVC at 53.2 Pa, (E) imidazole reacted to PVC at 26.6 Pa, and (F) imidazole reacted to PVC at 13.3 Pa.

be used to monitor the isocyanate groups at the film–air and film–substrate interfaces as a function of relative humidity and time. It is apparent that the concentration of isocyanate is lower at the film–substrate interface, and this difference can be attributed to the greater mobility of reactive functional groups near the film–substrate interface.

In many past ATR studies, surface depth profiling measurements were accomplished by varying the angle of incidence or by using ATR crystals with different optical densities. Among the early uses of ATR were the stud-

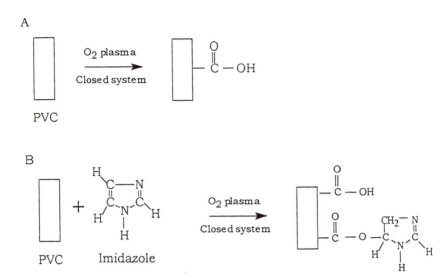

Figure 8.8. *Proposed mechanism of imidazole reaction with solvent-cast PVC surface in the presence of oxygen plasma.*

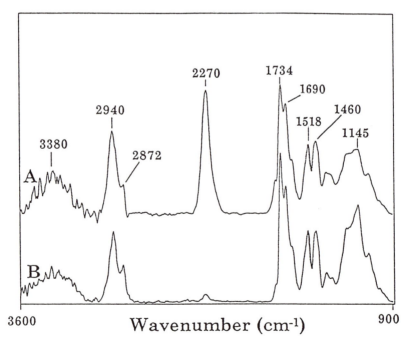

Figure 8.9. *ATR FT-IR spectra of urethane allowed to cross-link for (A) 1 day and (B) 80 days.*

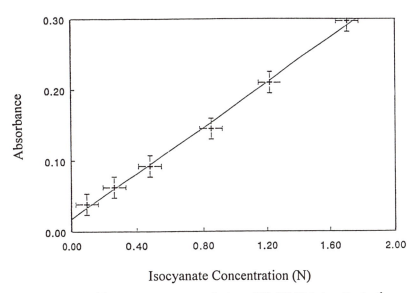

Figure 8.10. Calibration curve correlating ATR FT-IR intensity to the concentration of isocyanate functionality.

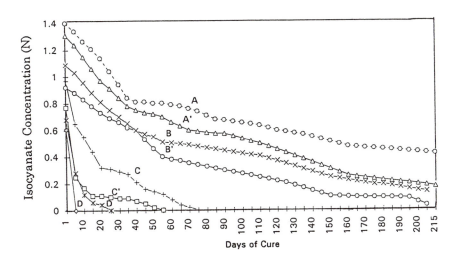

Figure 8.11. Concentration of isocyanate functionality at the film–air (F–A) and film–substrate (F–S) interfaces plotted as a function of relative humidity (RH) and time: (A) 0% RH, F–A; (A') 0% RH, F–S; (B) 20% RH, F–A; (B') 20% RH, F–S; (C) 40% RH, F–A; (C') 40% RH, F–S; (D) 80% RH, F–A; and (D') 80% RH, F–S.

ies of morphology and structure of polyurethanes at various depths (*24*). Subtraction of two spectra obtained at two different depths yielded a spectrum of the material representative of that region of the sample. Although this study showed that there are two distinctly different morphological forms of polyurethane, one on the surface and the other in the bulk of the polyurethane films, it should be kept in mind that the equation for the penetration depth (eq 2.8) was derived with an assumption that the surface is homogeneous; that is, there is only one species present. Jacobsen (*25*) has demonstrated a similar approach for obtaining intermediate layer spectra (solid and molten), by changing the angle of incidence. There are other examples of similar surface depth profiling (*26, 27*).

Polyester–Melamine

Most studies of polymeric films have explicitly dealt with either surfaces or the bulk. The issue of nonuniform distribution of species across a thermosetting polymeric film deposited on a substrate was addressed only in the early 1990s. The first studies addressing these issues were conducted on polyester–melamine films (*28*) and showed that the melamine distribution may be nonuniform. Figure 8.12 illustrates a series of ATR FT-IR spectra of linear polyester–melamine films cured at 250 °C for 1 min. The spectra were recorded from the film-air (traces A, B, and C) and film–substrate (traces D, E, and F) interfaces. Using the melamine band intensity at 815 cm^{-1} ratioed to the carbonyl band at 1725 cm^{-1} as a probe of the melamine concentration, the melamine content as a function of depth can be determined. As shown in Figure 8.13, the melamine content is higher near the film–air interface, probably because of melamine self-condensation. Self-condensation may be influenced by the amount of acid catalyst, the hydroxyl number of the polyester, the film-thickness, and the reaction rate differences between the film-air and film-substrate interfaces. In fact, this situation may not be uncommon in many other coating systems because a nonuniform distribution of film components was also observed for epoxy–polyamide (*29*) and melamine–polyol–phthalocyanine (*22*) thermosetting thin films.

Silicon-Containing Polymers

A comparison of the surface analysis of bisphenol A–polycarbonate/poly(dimethylsiloxane) random copolymers was illustrated by Mittlefehldt and Gardella (*30*) who used Fourier self-deconvolution (*31*) to resolve ATR spectra and compared these results with transmission measurements. The results of ATR analysis appear to be different from those obtained in transmission measurements using band deconvolution procedures. This exam-

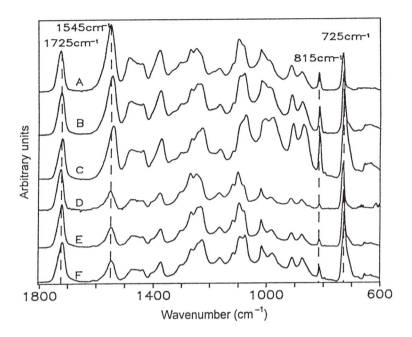

Figure 8.12. Series of ATR FT-IR spectra of linear polyester–melamine films cured at 250 °C for 1 min (0.5% w/w catalyst, film thickness 25 μm) obtained from various depths at the film–air (A, B, and C) and film–substrate (D, E, and F) interfaces. (Reproduced with permission from reference 28. Copyright 1992 Elsevier Sequoia.)

ple reaffirms that optical effects must be taken into account for adequate quantitative ATR analysis.

Poly(dimethylsiloxane) (PDMS) compounds are a group of cross-linked polymers with many useful properties, especially when chemical inertness, flexibility, and mechanical integrity are needed. For these reasons, they are used in many biomedical applications (*32, 33*). In an effort to make them more biocompatible, several studies were conducted on their surface modifications. Among many surface modification techniques, plasma surface reactions are attractive because they offer the ability to favorably alter a polymer surface without changing the structure and properties of the bulk polymer.

Silicones are particularly important because their derivatives are among the most commonly used elements in semiconductors. For that reason ATR as well as other surface approaches were used extensively to elucidate the origin of surface structures. However, ATR applications were limited, and reflection–absorption spectroscopy dominated the field. The silicon-related

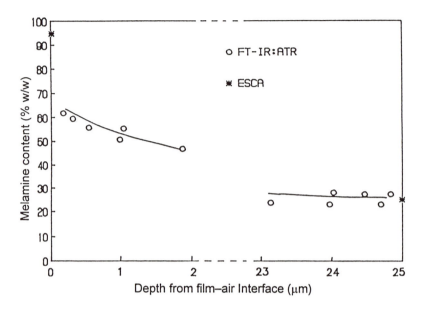

Figure 8.13. Depth profile of the melamine content for the linear polyester-melamine coating obtained using ATR FT-IR and ESCA measurements. (Reproduced with permission from reference 28. Copyright 1992 Elsevier Sequoia.)

studies go back to the 1960s, when Becker and Gobeli (*34*) detected the weak Si–H stretching modes resulting from H chemisorption on silicon, and to Beckmann and Harrick (*35*), who measured relative hydride and hydroxyl contents on silicon. More recent studies on semiconductors involve ultrahigh vacuum conditions or atmospheric pressure depositions after chemical treatments. In an excellent review, Chabal (*36*) presented several examples of how structures of various silicon-containing clusters, including mono-, di-, and trihydrides can be determined on rough or smooth silicon surfaces (Si[100], Si[111]). Surface etching using hydrogen fluoride was also described and further explored in later studies (*37, 38*) Other applications of internal spectroscopy include diffusion measurements of adsorbates on semiconductor surfaces and other metals—for example, CO on Pt[111] (*39, 40*)—laser-induced photochemical processes (*41*).

ATR was used for the analysis of PDMS and showed that the presence of chloro-functional molecules in PDMS inhibits surface functionalization when ammonia plasma is used (*42–45*). Using insert Ar and N_2 gases in microwave environments allows the formation of the Si–H surface entities that give two IR bands at 2158 and 912 cm^{-1} due to the Si–H stretching and bending modes, respectively (Figure 8.14).

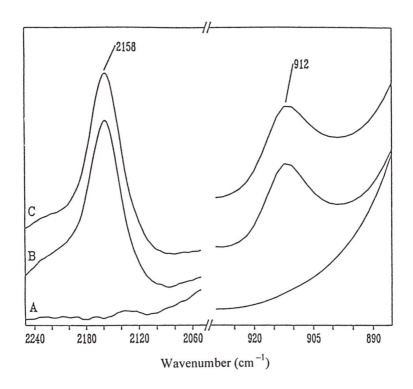

Wavenumber (cm^{-1})

Figure 8.14. ATR FT-IR spectra of poly(dimethylsiloxane) in the Si–H stretching and bending regions for (A) untreated, (B) N_2-plasma treated, and (C) Ar–plasma treated. (Reproduced with permission from reference 42. Copyright 1992 Butterworth–Heinemann Ltd.)

Figure 8.15 illustrates ATR FT-IR spectra in the C=C and C=N stretching regions for imidazole reacted to PDMS surface at 106.7, 53.3, and 26.6 Pa in a closed reactor (*46, 47*) using s-polarization (traces A, B, and C) and p-polarization (traces D, E, and F). These observations indicate that, under the given conditions, imidazole molecules react with the PDMS surface and also take a preferentially parallel orientation to the surface. One of the reaction mechanisms leading to the PDMS–imidazole reactions is illustrated in Figure 8.16.

Quantitative Analysis Example I

An inherent drawback of ATR measurements—and, for that matter, IR spectroscopy—is the lack of an internal calibration method. Therefore, the first step is to construct a calibration curve. For Si–H species created as a result of microwave plasma reactions, as illustrated in Figure 8.14, this can be

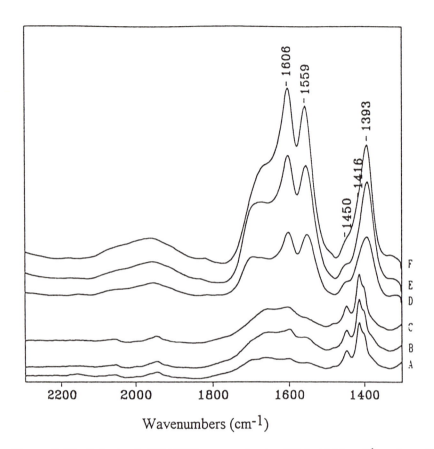

Wavenumbers (cm⁻¹)

Figure 8.15. Polarized ATR FT-IR spectra in the 2300–1300 cm⁻¹ region of imidazole reacted to PDMS surface for 10 s at various pressures using closed reactor: (A) s-polarized spectrum at 106.7, (B) s-polarized spectrum at 53.3 Pa, (C) s-polarized spectrum at 26.6 Pa, (D) p-polarized spectrum at 106.7 Pa, (E) p-polarized spectrum at 53.3 Pa, and (F) p-polarized spectrum at 26.6 Pa. (Reproduced from reference 46. Copyright 1995 American Chemical Society.)

achieved by measuring the ATR spectra of homogeneous liquid standards containing 0.1 to 0.35 M bis(trimethylsiloxy)methylsilane (BTMS) and plotting intensity as a function of concentration. For a solvent, liquid PDMS can be used because it has no absorption bands in the Si–H stretching region and because, being a polymer, it will resemble the environments of the cross-linked networks. Now, if the ATR band intensity plotted as a function of concentration is independent of the optical effects, for each angle of incidence, an identical straight line should be obtained. However, for the

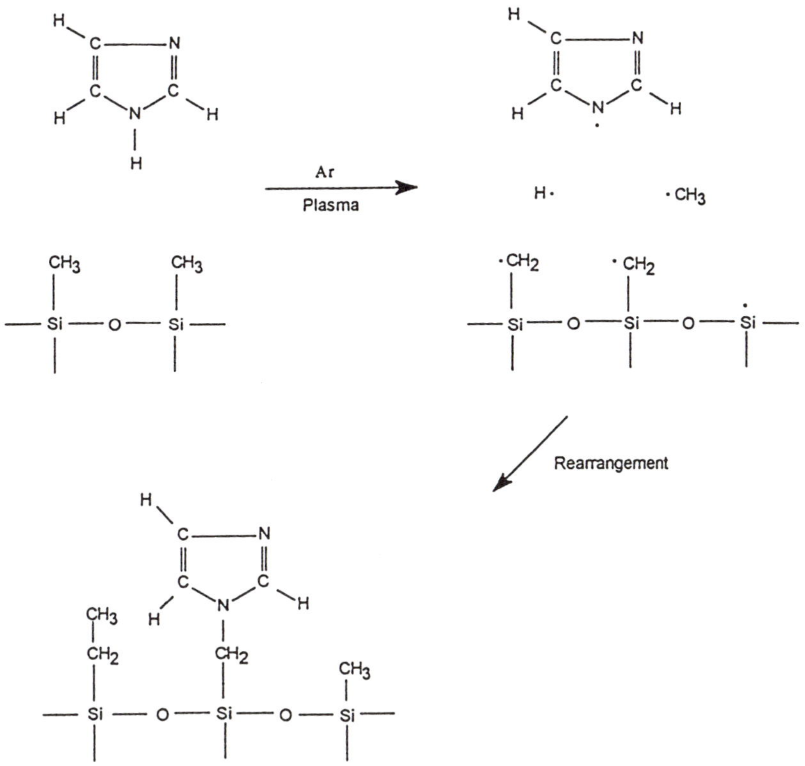

Figure 8.16. Reaction mechanism between imidazole and PDMS surface in Ar microwave plasma. (Reproduced from reference 47. Copyright 1996 American Chemical Society.)

raw data illustrated in Figure 8.17, this is not the case. Therefore, the calibration curves in Figure 8.17 are of no use in determining Si–H concentrations because the slopes change with concentration. The data illustrated in Figure 8.16 also indicate that, for the standards with a common refractive index and the same surface coverage, the band intensities increase with decreasing angle of incidence. This is expected because the spectra obtained at greater angles of incidence contain information from thinner surface layers and fewer reflections. This results from a smaller effective sample thickness being analyzed at larger angles. However, when the raw ATR spectra are subjected to the algorithm that accounts for optical effects (Figure 6.5), the plots should be linear and superimposed. Indeed, when ATR spectra are corrected, linear and superimposed calibration curves illustrated in Figure 8.18 are obtained. Only slight deviations are detected for a 60° incidence beam; these can be attributed to a random polarization of

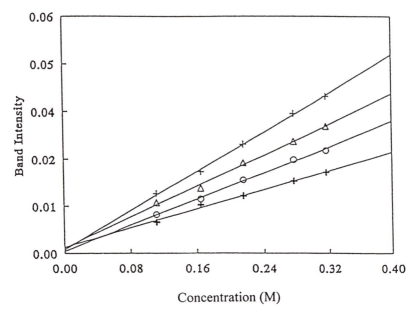

Figure 8.17. Concentration as a function of Si–H band intensity for bis(tri-methylsiloxy)methylsilane (BTMS) standard solutions recorded at various angles of incidence: +, 45°; Δ, 50°; O, 55°; +, 60°. (Reproduced with permission from reference 42. Copyright 1992 Butterworth–Heinemann Ltd.)

light at this angle of incidence. This effect commonly increased as the angle of incidence deviates from the angle that allows light to pass perpendicularly into the crystal surface. Each line in Figure 8.18 has a linear correlation coefficient greater than 0.996. Thus, the calibration curve is suitable for determining the Si–H concentrations.

However, each analysis of that nature should be tested and estimated for accuracy. After all, these calibration curves will be used for estimating the unknown amounts. One approach is to use species containing Si–H in distinctly different chemical environments. In this case, we can choose methylhydro–dimethylsiloxane copolymer, make standard solutions with known concentrations, and compare the concentrations with those determined from a calibration curve. Table 8.1 provides data obtained for a 0.1 M solution; the results show good precision, with a deviation of 0.01 M between four calculated values from recorded spectra at various angles of incidence. One important experimental issue that should be remembered is that the plots in Figure 8.16 were obtained using species containing Si–H bonds, and their environment does not seem to influence the extinction coefficients. This observation may not hold for other systems.

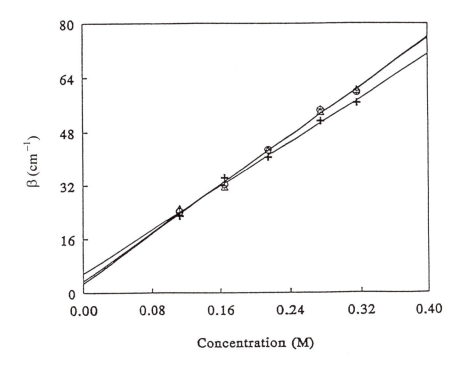

Figure 8.18. Plot of BTMS concentrations as function of Si–H band intensity after Urban–Huang algorithm recorded under the same conditions as listed for Figure 8.9. (Reproduced with permission from reference 42. Copyright 1992 Butterworth–Heinemann Ltd.)

Table 8.1. Calculated from ATR data and actual concentrations of Si–H groups in poly(dimethylsiloxane)

Effective angle (°)	β (cm⁻¹)	Actual Si–H concentration (M)	Calculated Si–H concentration (M)
45.00	21.18	0.10	0.08
47.10	22.18	0.10	0.08
49.28	22.96	0.10	0.09
51.24	24.78	0.10	0.09

SOURCE: Reproduced with permission from reference 42. Copyright 1992 Butterworth–Heinemann Ltd.

If use of the algorithm described in Chapter 6 is indeed necessary for quantitative analysis, the calibrations illustrated in Figure 8.18 should not be affected by the ATR crystal surface coverage. To test this hypothesis, ATR spectra can be recorded for a known, standard sample with the reduced crystal coverage. The results obtained for Si–H stretching vibrations of methylhydro-dimethylsiloxane presented in Table 8.2 indicate that for a 35% reduction of the surface coverage and an actual concentration of 0.21 M, the calculated concentrations from the ATR data are not affected.

Since several tests were successfully passed, we can now use the calibration curve shown in Figure 8.18 to determine how the Si–H concentration changes as a result of microwave plasma at various depths of penetrations. To do so, we change an angle of incidence, and the data are presented in Table 8.3. Going from 0.6 to 0.8 μm into the surface, the concentration of Si–H changes by about 25%. Keep in mind, however, that the concentrations calculated in Table 8.3 do not represent local concentration at a given specific depth but are average concentrations for the entire depth of penetration. To obtain local concentrations, we would need to mechanically strip the surface or use the nondestructive stepwise stratified theory outlined in Chapter 7.

Table 8.2. Calculated from ATR data using reduced surface coverage (about 35%) and actual concentrations of Si–H groups in poly(dimethylsiloxane)

Effective angle (°)	β (cm^{-1})	Actual Si–H concentration (M)	Calculated Si–H concentration (M)
45.00	45.28	0.21	0.20
47.10	45.09	0.21	0.20
49.28	46.40	0.21	0.21
51.24	46.41	0.21	0.21

SOURCE: Reproduced with permission from reference 42. Copyright 1992 Butterworth–Heinemann Ltd.

Table 8.3. Si–H concentration changes as a result of PDMS microwave plasma treatment as a function of the depth of penetration

Effective angle (°)	β (cm^{-1})	Depth of penetration (μm)	Calculated Si–H concentration (M)
45.00	31.04	0.807	0.15
47.10	31.76	0.719	0.16
49.18	35.24	0.657	0.18
51.24	37.36	0.611	0.19

SOURCE: Reproduced with permission from reference 42. Copyright 1992 Butterworth–Heinemann Ltd.

It is useful to modify surfaces in such a way that the species reacted to the surface have groups selective and specific for other surface reactions. It is also desirable to use monomers that can be bonded to surfaces. For that reason, Gaboury and Urban (48) examined how two pairs of monomers—succinic anhydride and propionamide, along with their unsaturated counterparts, maleic anhydride and acrylamide—can be deposited by microwave plasma and quantitatively analyzed. Figure 8.19 shows a series of ATR spectra recorded on succinic (A) and maleic (B) anhydride modified PDMS surfaces; Figure 8.20 shows the spectra of propionamide (A) and acrylamide (B) modified PDMS surfaces. It is apparent that the presence of the C=C double bond is necessary for such species to be chemically bonded to the surface, and depending on the microwave plasma exposure time, various surface species illustrated in Figure 8.21 may predominate. Although the advocates of X-ray photoelectron spectroscopy measurements could argue that the reactions of species not containing C=C bonds can also be detected, we should realize that any high-energy microwave exposure will modify a surface somehow; even inert gases will modify surfaces, but the reproducibility of functionalities and their usefulness may be another issue.

Quantitative Analysis Example 2

Let us use the data presented in Figures 8.19B and 8.20B and set up a method for quantitative surface analysis using ATR. First, we will identify experimental procedures necessary for such analysis. ATR spectra with an adequate signal-to-noise ratio should be recorded using polarized light. A typical spectrometer resolution of 4 cm^{-1} is suitable for such studies. An ATR cell can be aligned at a 60° angle of incidence with a 45° end parallelogram KRS-5 crystal (it is essential to remember that the angle of incidence should not be too close to the critical angle; see Chapter 4). The spectral acquisition time may vary from case to case, but 60 to 400 scans are usually sufficient. The reference (empty cell) and sample spectra should be recorded under the same alignment and purging conditions* using the same polarizer (perpendicular polarization using an aluminum wire grid polarizer with a 0.4 μm grid spacing is perfectly suitable). Once the spectra are collected, they should be converted using the algorithm illustrated in Figure 6.5 to a form independent of the sample refractive index, crystal surface coverage, and angle of incidence, thus allowing the use of the Beer–Lambert law. For determination of extinction coefficients, several concentration standards should be prepared, and their transmission spectra should be recorded. For example, Figure 8.22 is a plot of absorbance versus con-

---

*All infrared instruments are purged to eliminate water and carbon dioxide, since these species absorb strongly in the infrared region. Sufficient purge is accomplished by using molecular nitrogen (an expensive way to keep an instrument operational), special filters, and compressed air.

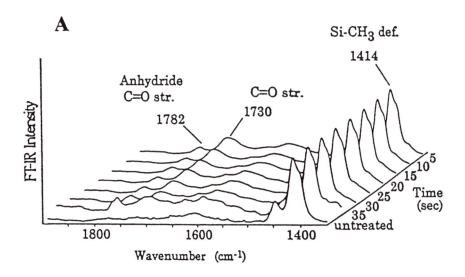

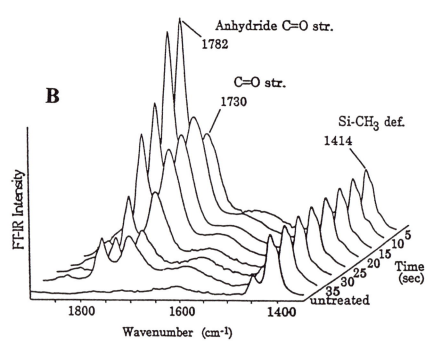

Figure 8.19. ATR FT-IR spectra of (A) succinic and (B) maleic anhydride plasma modified PDMS plotted as a function of time. (Reproduced with permission from reference 44. Copyright 1994 American Chemical Society.)

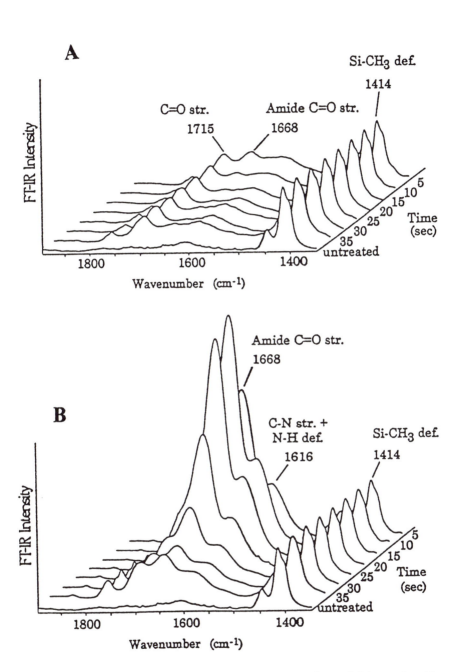

Figure 8.20. ATR FT-IR spectra of (A) propionamide and (B) acrylamide plasma modified PDMS plotted as a function of time. (Reproduced with permission from reference 44. Copyright 1994 American Chemical Society.)

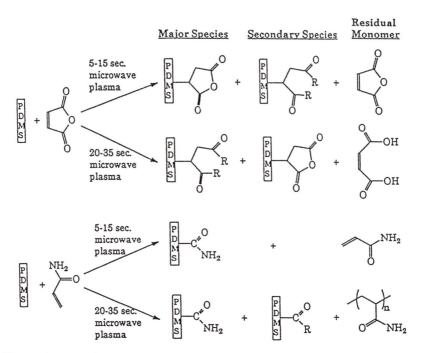

Figure 8.21. Schematic diagram of microwave-plasma-induced reactions on a silicone elastomer (PDMS) surface. (Reproduced from reference 44. Copyright 1993 American Chemical Society.)

centration for maleic anhydride and acrylamide solutions. The respective extinction coefficients, ε, are 1148 and 563 L/(mol cm). Once the spectra are corrected, concentrations can be calculated from the Beer–Lambert law ($\beta = \varepsilon c_2$). The β values are obtained from the corrected spectra. Tables 8.4 and 8.5 list average β values, volume concentrations, and surface concentrations for maleic anhydride and acrylamide spectra from Figures 8.19 and 8.20, with the reaction time. Again, it should be realized that the major source of error is in the measurement of extinction coefficients; the extinction coefficients were determined on maleic anhydride and acrylamide solutions using transmission measurements. If chemical bonding to the surface has a significant effect on the extinction coefficients, such an approach may not be advisable.

Latexes

Before we examine possible applications of ATR spectroscopy in latex analysis, note that the concept of distribution of the low molecular weight species in polymer matrices may have significant implications not only for diffusion but also for other properties such as adhesion or durability. While

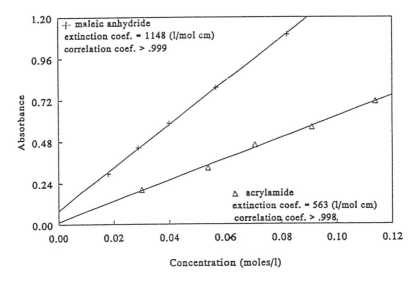

Figure 8.22. Plots of C=O stretching band absorbances for maleic anhydride and acrylamide as a function of concentration. (Reproduced from reference 45. Copyright 1994 American Chemical Society.)

Table 8.4. Average β Values, volume concentrations, and surface concentrations for acrylamide-plasma-treated PDMS

	Reaction time (s)	Avg. β at 1782 cm^{-1} (cm^{-1})	Volume conc. (10^{-8} mol/cm^3)	Surface conc. (10^{-12} mol/cm^2)
Maleic anhydride	5	0.0265 ± 0.0070	2.31 ± 0.61	1.71 ± 0.45
	10	0.0256 ± 0.0086	2.23 ± 0.75	1.65 ± 0.55
	15	0.0195 ± 0.0073	1.70 ± 0.64	1.25 ± 0.47
	20	0.0183 ± 0.0049	1.59 ± 0.42	1.18 ± 0.31
	25	0.0183 ± 0.0087	1.60 ± 0.76	1.18 ± 0.56
	30	0.0161 ± 0.0016	1.41 ± 0.14	1.04 ± 0.10
	35	0.0076 ± 0.0005	0.66 ± 0.04	0.49 ± 0.03

SOURCE: Reproduced with permission from reference 45. Copyright 1992 Butterworth–Heinemann Ltd.

introducing low molecular weight species into polymer networks is a well-known and common method for improving numerous properties, the question of how their behavior may affect such properties as adhesion or durability was not addressed until early 1990s. For example, plasticizers are often introduced into a polymer matrix in an effort to lower the temperature

Table 8.5. Average β Values, volume concentrations, and surface concentrations for anhydride-plasma-treated PDMS

	Reaction time (s)	Avg. β at 1665 cm^{-1} (cm^{-1})	Volume conc. (10^{-8} mol/cm^3)	Surface conc. (10^{-12} mol/cm^2)
Acrylamide	5	0.0204 ± 0.0071	3.61 ± 1.26	2.86 ± 1.00
	10	0.0312 ± 0.0058	5.53 ± 1.03	4.38 ± 0.82
	15	0.0286 ± 0.0052	5.07 ± 0.93	4.02 ± 0.73
	20	0.0246 ± 0.0054	4.36 ± 0.97	3.45 ± 0.76
	25	0.0096 ± 0.0018	1.70 ± 0.32	1.35 ± 0.26
	30	0.0085 ± 0.0037	1.41 ± 0.66	1.19 ± 0.53
	35	0.0049 ± 0.0010	0.88 ± 0.18	0.69 ± 0.14

SOURCE: Reproduced with permission from reference 45. Copyright 1992 Butterworth–Heinemann Ltd.

of processibility by effectively lowering the overall T_g of the system. These monomeric species may, however, significantly influence adhesion to plastic substrates if they migrate to the film–air interface. While the use of plasticizers may be avoided, in latex systems the presence of surfactants during the latex preparation is a necessity governed by the formation of surfactant micelles that allow the latex particle formation.

Synthesis of polymers by emulsion polymerization appears to be a highly attractive method because it results in a colloidal dispersion of polymer or copolymer latex particles in an aqueous medium. Such dispersions coalesce to form continuous polymer films, and many useful synthetic routes for latex preparation are now well documented. Along with the practical aspects of latex synthesis, several debatable theories concerning the mechanisms of coalescence and the processes associated with it have been proposed. However, the issues of physicochemical interactions between individual latex components and the film interfacial properties resulting from these interactions (48–50) are particularly pertinent in a context of polymer or copolymer composition, particle composition, and surfactant compatibility. It is, therefore, appropriate to address the question of how the presence of unavoidable surfactants may influence latex film properties, particularly adhesion. If surfactant molecules are incompatible with the polymer latex network after latex coalescence, we may anticipate a phase separation, diffusion, or enhanced mobility, as these small molecules are free to move within the network. Furthermore, if flocculation occurs prior to coalescence, surfactant molecules may be displaced from the latex particle surfaces as a result of two particles coming into contact. If, after coalescence, such surfactant molecules are capable of migrating in some preferential direction across the film, the direction of propagation will affect such film properties as adhesion and durability. This is because most surfactants are water-soluble and, if their concentration at the film–substrate interface is increased, they may be washed away, and their loss can lead to adhesion failure. One of the puzzling observations made using ATR FT-IR spectroscopy was the possibility of monitoring surfactant exudation to the film–air or film–substrate interfaces. For example, Figure 8.23 illustrates ATR FT-IR spectra recorded from the film–air and film–substrate interfaces. It appears that the bands at 1046 and 1056 cm^{-1}, which were attributed to the S–O stretching modes of the hydrophilic end of a surfactant molecule, detected only at the film–air interface. Furthermore, the

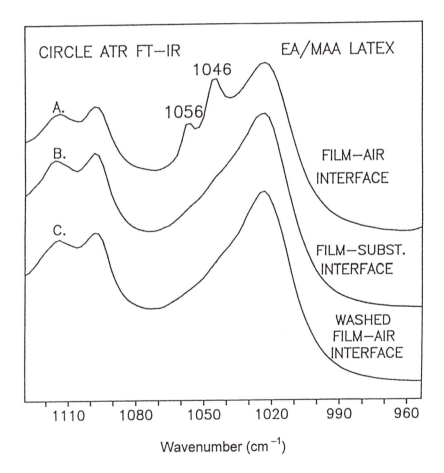

Figure 8.23. ATR FT-IR spectra of butyl acetate/methacrylic acid latex: (A) film–air interface, (B) film–substrate interface, and (C) film–air interface washed with methanol–water solution. (Reproduced with permission from reference 51. Copyright 1992 Butterworth–Heinemann Ltd.)

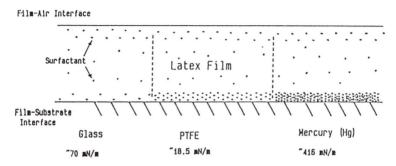

Figure 8.24. Schematic representation of the substrate effect on the distribution of surfactant molecules across the latex films prepared on glass, poly(tetrafluoroethylene), and liquid mercury. (Reproduced with permission from reference 49. Copyright 1991 Plenum.)

bands disappear when the surface is washed with water. The direction of exudation was found to depend on the surface tension of the substrate and the external forces imposed on the film. Figure 8.24 illustrates a particular case of how the surface tension of a substrate may affect the distribution of sodium dioctyl-sulfosuccinate (SDOSS) surfactant molecules across the ethyl acrylate/methacrylic acid (EA/MAA) latex. Apparently, the distribution depends on the water flux out of the film in the early stages of film formation and the surface tension of the substrate. In the most recent studies, the effects of interfacial orientation of the latex copolymer acid groups (which, under suitable conditions, can form dimers) and the hydrophilic groups of the surfactants were also examined (*52, 53*).

Another study focused on monitoring molecular processes and interactions that are responsible for dynamics, mobility, orientation, and the effect of external sources on surfactant molecules in EA/MAA coalesced latex films (*54–56*). Polarized ATR FT-IR spectroscopy, as described in Chapter 2, was used to identify the mobility and surfactant exudation of SDOSS surfactant molecules to the film–air and film–substrate interfaces in styrene/*n*-butyl acrylate latex films (*57*). It was established that, depending upon the latex particle composition, the surfactant molecules can be driven to the film–air or film–substrate interfaces. The primary factors that govern the direction of exudation are the compatibility of the latex components, interfacial film–substrate surface tension, and the chemical composition of the latex particles. Concentration as well as orientation of the hydrophilic $SO_3^-Na^+$ surfactant ends change as a function of depth and latex particle composition. Figure 8.24 illustrates how the concentration of SDOSS changes at film-air and film-substrate interfaces as a function of depth. These data demonstrate that the SDOSS concentration diminishes rapidly at shallow depths but levels off at around 1.7 μm from either interface. In a similar study,

Fina et al. (*58*) examined the factors leading to Triton-100 surfactant migration toward or away from the film–air interface. Apparently, migration of this surfactant is independent of its concentration in the aqueous formulation but depends on drying and aging conditions.

Multicomponent Systems

Although there is already a significant effort devoted to multicomponent systems, future research will focus more on hybrid materials and their blends, in which interfacial interactions, stability, and mutual compatibility are of primary importance. Therefore there will be an increasing need for more sensitive techniques that allow detection of minute amounts of interfacial or surface species. ATR spectroscopy is not only readily accessible and sensitive, but with new developments in fiber optics, sensitive detectors, and proper spectral analysis, it will remain an attractive technique in many areas of research, ranging from determining optical constants of substances in various regions of the electromagnetic spectrum to qualitative and quantitative analysis of surfaces and interfaces. The following sections provide examples in which form of ATR spectroscopy was used to solve various problems.

Particle–Polymer or Fiber–Polymer Interactions

As a starting point, let us realize that although quantitative characterization of the interfacial regions between metal, ceramic, or polymeric surfaces has long been a goal of science, there are still many issues to be sorted out. The primary problems are adherence between the two substances which is usually associated with intrusion of water (*59*); subsequent electrochemical processes at the interface (*60*), often followed by crack formation and propagation due to internal stresses and thermal cycling; and simple incompatibility between substances.

Carbon black is used not only as a filler for elastomers but also in other applications. Interest in its surface modifications goes back to the early 1960s, but spectroscopic studies were not initiated until the 1970s (*61*). Figure 8.25 is an internal reflection spectrum of air-oxidized carbon black obtained by pressing the fiber against a 45° incidence Ge crystal (*62*). The spectrum shows that despite opacity of the sample, it is still possible to obtain good-quality spectra. Spectroscopically, a similar sampling situation can be encountered when studying carbon or graphite fibers. Studies by Ishino and Ishida (*63*) on oxidized graphitized carbon fiber interfaces clearly showed that, along with experimental ATR advances demonstrating that the presence of surface acid groups can activate interfacial cross-linking reactions, a theoretical optical approach can be applied to quantify the results.

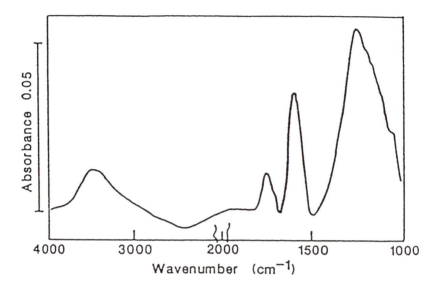

Figure 8.25. ATR spectrum of an air-oxidized carbon black. (Reproduced with permission from reference 62. Copyright 1987 Chapman and Hall Ltd.)

Surface oxidation results in four main bands due to carbonyl (1720 cm^{-1}), quinone (1580 cm^{-1}), and the C–O stretching and O–H bending modes (both at approximately 1200 cm^{-1}).

Several model studies on composite interfaces were carried out for carbon and graphite fibers (*64, 65*). Because graphite exhibits semiconductivity and a high absorption coefficient, it can enhance the IR bands of a thin film of poly(vinyl acetate) (*66, 67*). Because graphite fibers have a variety of uses, ranging from fibers in composites to pigmentation and fillers in coatings, an interesting spectroscopic approach was taken by Claybourn et al. (*68*). It involves the use of specular reflection spectroscopy at normal incidence and uses an IR microscope to collect the spectra, which are converted to absorbance by KKT. This study, as well as other recent studies on carbon-filled polymers, disproves previous beliefs that surface analysis of carbon black by IR transmission was made difficult by the nonreflecting, highly absorbing nature of the material. Other applications include studies of influence of carbon fibers (*69, 70*) and aramid fibers (*71*) on epoxy curing using thin films of carbonized polyacrylonitrile and aramid.

One of the useful ATR applications for fibers is the circular crystal. In essence, a fiber or polymeric film sample is wound tightly on an ATR crystal and the spectra are recorded. This approach can be used to obtain ATR spectra of aramid fibers and polymeric films (*72*).

The effects of fillers other than carbon black on polymeric systems can be also examined by ATR (*73*). For example, when two components to be reacted are mixed in proper ratios—say, a diglycidolether hydrogenated bisphenol-A (DGEHPA) based epoxy cross-linked with a polyamide based on a fatty acid dimer and triethylenetetramine, and pigment or inorganic filler particles are added, it is likely that adsorption of the particle surfaces of one of the cross-linking components will occur. As a result of such adsorption, the stoichiometric ratio of the reactants will change, thus preventing the formation of a fully cross-linked network because a fraction of the unreacted component may simply be dispersed in the network. Figure 8.26 shows two ATR spectra: one recorded at the film–air interface and the other at the film–substrate interface. A comparison of the spectra indicates that the film–air interface contains an excess of polyamide species, as demonstrated by the 1605 cm^{-1} band. Another scenario that sometimes occurs is that one of the components may have a tendency to migrate to either film–air or film–substrate interfaces. In this case it is helpful to consider the surface tensions of individual components; for example, polyamide exhibits lower surface tension (about 35 dyn/cm) than its epoxy counterpart (about 50 dyn/cm) (74), and therefore it may preferentially migrate to the surface to compensate for the excess of surface energy. We might also anticipate wavenumber shifts as a result of environmental differences between the surface and the bulk.

Polymer–Metal Interfaces

Because the structure of the polymer–metal interface is one of the critical factors that governs adhesion, and subsequently corrosion resistance, chemical bonding across the interface is considered to be of greatest importance. Although there are other, perhaps preferred, approaches to elucidating the origin of interfacial chemistry, ATR spectroscopy has also been applied. One example is the degradation of polyethylene on copper (75, 76), which exhibits long-range interfacial effects. An interesting interfacial phenomenon was also observed in polybutadiene, which oxidizes catalytically in the presence of certain metal oxides (77). Curing on metal surfaces results in the formation of an interfacial region of polymer that is more oxidized than the bulk (78, 79).

Water at the adsorbate–metal interface is one of the most significant factors affecting polymer–substrate interactions, and the water entering the adsorbate environment may result in electrochemical reactions at the metal surface that lead to corrosion. Although the FT-IR spectrum of water contains two major bands that are easily identified—O–H stretching (3400 cm^{-1} region) and the bending mode at about 1640 cm^{-1}—the measurement of orientation or rearrangements resulting from intrusive behavior of water is not a simple matter. The complication is often related to the presence of bands in the 3400 cm^{-1} region. These bands consist of two or more sepa-

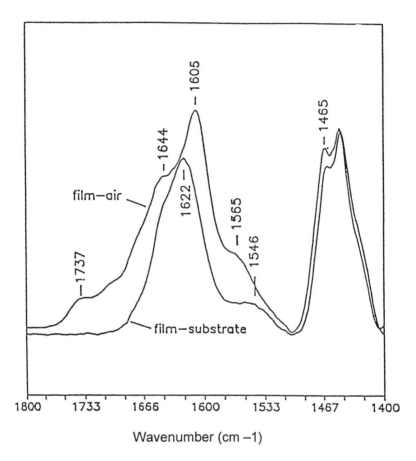

Figure 8.26. ATR FT-IR spectra of film–air and film–substrate interfaces of pigmented coating based on hydrogenated bisphenol-A-based epoxy resins cross-linked with polyamide based on fatty acid dimer and triethylene-tetramine. (Reproduced with permission from reference 73. Copyright 1993 John Wiley & Sons.)

rate spectral features, and the situation becomes even more complex when water molecules are adsorbed on the surface (*80*). Nguyen et al. (*81*) have used in situ ATR FT-IR spectroscopy for monitoring diffusion of water through pigmented and alkyd coatings. As a result of diffusion, interfacial degradation was detected. Besides the expected intensity increase of water bands and diminishing bands due to epoxy or alkyds, no other changes were detected.

If the film–substrate interactions result in the formation of water-soluble species, an interfacial failure will occur. Such interfacial properties often develop if the substrate surface is pretreated with zinc-galvanized steel. Interfacial failure can also occur if polymeric coatings deposited on the substrate are zinc-enriched (*82*). Makishima et al. (*83*) have conducted experiments in which zinc- and tin-galvanized steel samples were coated with long-chain alkyd paints and then exposed outdoors. They analyzed interfacial products by peeling off the coating and analyzing the interface by ATR spectroscopy. It was found that the formation of zinc soaps and oxidation products at the coating–metal interface is responsible for the substantial decrease of adhesion. They also determined the effect of Ca_2PbO_4 added to the paint formulation and found it to significantly enhance adhesion.

O'Brien and Hartman (*84*) used ATR spectroscopy to interpret the mechanism of paint adhesion to wood by using a cellulose/epoxy resin model system. An IR spectrum of cured epoxy resin on cellulose was obtained by placing a cast epoxy resin sheet on one side of the crystal and a dry, pressed, reconstituted cellulose substrate on the other side. The recorded spectrum was compared with that of a cast cellulose/epoxy mixture and cellulose paper impregnated with epoxy resin. The analysis showed that the characteristic bands of cellulose were heavily diminished and the 775 cm^{-1} band due to epoxy was absent. These studies allowed the determination of chemical reactions responsible for adhesion.

Adhesion

The apparent lack of polar groups on the surfaces of polymeric substrates is reflected in the difficulty of wetting these substrates, which subsequently inhibits attaching coatings. Hence, in an effort to improve adhesion, the surfaces are physically or chemically modified. Quite often, depending on the mode of application, the surfaces are subjected to plasma or corona discharge treatments in the presence of various gases. The choice of treatment is dictated by the chemical composition of the substrate or suitability of the particular treatment. One of the studied systems was gas–plasma modifications of silicone elastomers (*85, 86*). A combination of spectroscopic measurements and dynamic mechanical analysis provides a suitable approach to establish that among plasma surface treatments, only ammonia–plasma leads to an increase of the ln E' modulus values (E' is the storage modulus), whereas argon and carbon dioxide treatments produce oxidative products that decrease the storage modulus above the glass transition temperature T_g. These studies have also shown that previous findings (*87*) that the surface cross-link density increases as a result of gas–plasma treatments are not necessarily valid. On a similar note, two polyurethane substrates with different values of Young's modulus will exhibit different radii of contact with glass beads (*88*). This observation can be attributed to differences in the surface interactions between soft and hard segments of polyurethanes with glass. In other studies (*42*), it was shown that using microwave gas–plasma, silicone elastomers could be modified, leading to formation of Si–H functional groups. It was also found (*89*) that the presence of residual chlorine-containing monomers may impair gas–plasma surface modifications through formation of ammonium chloride.

In an effort to improve adhesion between metal and polymer surfaces, it is common to modify the surface of the polymer by either acid etching or corona discharge treatments. In addition, it is necessary to remove traces of impurities on a metal surface in order to achieve desirable interfacial bonding. For example, surface contamination in the processing of aluminum sheets is a severe problem. Moreover, the presence of water on the surface of aluminum often leads to the conversion of amorphous Al_2O_3 to hydrous oxides such as AlOOH (boehmite) or $Al(OH)_3$ (bayerite) (*90–92*). To prevent these processes, sodium heptagluconate and phosphoric acids are often used. While chemical treatments are used to improve bonding between the aluminum surface and polymers, physical treatments may also enhance this process. Such procedures as roughening of aluminum surfaces are employed, along with various etching treatments (sulfuric acid or sodium dichromate, phosphoric or chromic acid anodization) (*93*) in an effort to enhance mechanical interlocking with the overlaid polymer. Phosphoric acid anodization treatments additionally lead to the formation of an insoluble $AlPO_4$ monolayer that diminishes the rate of conversion of Al_2O_3 to $Al(OH)_3$. The effect of surface modifications of polyethylene and aluminum on interlaminar adhesion was investigated (*94*). Both electron spectroscopy for chemical analysis (ESCA) and multiple internal reflection FT-IR spectroscopy measurements indicated the presence of hydroxyl, ester, and carbonyl groups on the $KMnO_4/H_2SO_4$-treated polyethylene surfaces. The presence of sulfonate and sulfate groups in localized areas of crevices was detected along with the surface vinyl species. Although most of these species were detected earlier by ESCA (*95*), the authors attributed the band at 1710 cm^{-1} to the COOH species and confirmed this assignment by the presence of C–O stretching vibrations in the $1260–1000$ cm^{-1} region as well as the C–H bending modes in the $1180–1085$ cm^{-1} region. While these assignments may indeed suggest the appearance of the COOH surface species, it is possible that tin salt in the presence of water-soluble cations will result in formation of the COO^-–cation pairs, giving rise to the symmetric and asymmetric C–O stretching vibrations that would absorb in the $1550–1400$ and $1350–1200$ cm^{-1} region (*96, 97*). Regardless of data interpretation and the mechanisms responsible for bonding, the adhesion between polyethylene and aluminum can be improved by modification of either surface, as was demonstrated by the T-peel adhesion test. Apparently, the oxidation of the polyethylene surface as well as incorporation of divalent cations before lamination enhance adhesion. Modifications of the aluminum surface with a double-bond-containing titanate lead to a threefold increase in adhesion, and upon mechanical testing, the failure occurs in polyethylene (*98*).

It seems fair to state that ATR is the least desirable technique for examining powders and other materials with high surface areas, primarily because of poor contact between the sample and the crystal. However, when inorganic and organic powders are overlaid by nondissolving liquids, the spectra of wetted powders are enhanced compared with the spectra of dry powders (*99*). Several examples of liquids such as water, acetone, deuterium oxide, or 2-propanol overlaying barium carbonate and other mineral powders were used to demonstrate that this approach may be of use in the analysis of powders.

Orientation of Macromolecules near Surfaces

When a polymer specimen is drawn to form a film, a certain orientation, usually parallel to a draw direction, may be induced, and certain macromolecular entities may take preferred orientations. Polarized light should be used to measure orientation changes, since it allows us to change the orientation of the electric vector of the incident radiation with respect to the laboratory axis and the polymer chain axis. There is another reason for using polarized light: every time the passing light strikes a mirror, 2% to 4% of the light will be polarized. As a matter of fact, even incident light is polarized because of Crawford's S factor (*100*). When using an ATR cell equipped with a rectangular internal reflectance element, the effect is further amplified, especially for surfaces with preferred molecular orientation, and cannot be adjusted for by using nonpolarized light (*101*).

With these remarks in mind, let us set the stage by considering a few examples of polarized transmission and ATR measurements and then outline theoretical foundations governing orientation measurements for ATR.

The experiment requires the use of an experimental setup illustrated in Figure 3.2A or 3.2B. Although for amorphous polymers we would anticipate no spectral differences between p-polarized and s-polarized vectors, external forces—for example, elongations, electromagnetic fields, or surfaces—may induce orientational changes. While it is relatively easy to envision orientation changes induced by elongating an amorphous polymer, or liquid crystalline polymers taking a certain orientation when electric fields are present, the presence of a surface induces special circumstances. When a molecule contains hydrophobic and hydrophilic groups, it will take a preferential orientation if a surface to which it is attached exhibits hydrophobic or hydrophilic properties. Such observations were possible with polarized ATR FT-IR spectroscopy, in which orientation changes of surfactant and acid molecules in EA/MAA and butyl acrylate/styrene latexes were examined (50, 57). Because these studies were strictly quantitative, we will use a polyethylene as an example.

First, let us see the difference between transmission and ATR polarized measurements. Figure 8.27 illustrates the CH_2 bending (1474 and 1464 cm^{-1}) and rocking (730 and 720 cm^{-1}) regions of polyethylene spectra. Trace A was recorded in transmission mode with the IR incident light polarized perpendicular to the machine direction of the sample, whereas for trace B the polarization was in the parallel direction. Traces C and D of Figure 8.27 are the same spectra recorded in the ATR mode at a 45° angle of incidence.

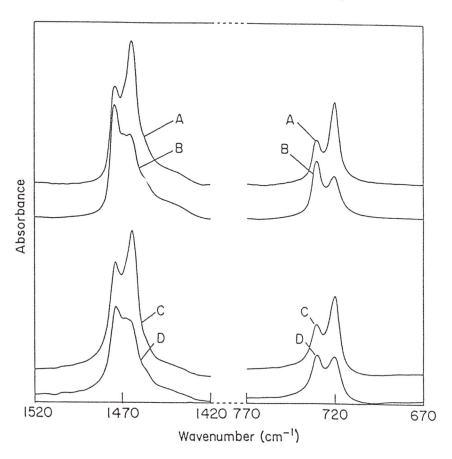

Figure 8.27. FT-IR spectra of polyethylene: (A) transmission perpendicular polarization, (B) transmission parallel polarization, (C) corrected ATR perpendicular polarization, and (D) corrected ATR parallel polarization. (Reproduced with permission from reference 8. Copyright 1992 Butterworth–Heinemann Ltd.)

First, let us compare transmission spectra illustrated by traces A and B. It appears that the intensity ratios of the 730 and 720 cm^{-1} bands are inverted when going from perpendicular (A) to parallel (B) orientations; that is, $(I_a/I_b)(I_a/I_b)_\perp < (I_a/I_b)_\parallel$. Although a similar trend is observed for ATR measurements (traces C and D), the changes are not as pronounced. In either case, the intensity ratio is sensitive to the orientation effect, but this sensitivity is diminished in a surface-sensitive ATR experiment because of a higher content of the randomly oriented amorphous phase near the surface; this phase has a glass transition temperature T_g below the temperature of experiment and therefore is randomly distributed throughout the film.

Let us further investigate the issue of orientation. Typically in orientation measurements, a so-called orientation function should be determined. This function gives a measure of the polymer chain orientation and is defined as

$$f = \left(\frac{D-1}{D+2}\right)\left(\frac{D_0+2}{D_0-1}\right) \tag{8.2}$$

where $D_0 = \cot^2\alpha$ (α is the angle between the transition moment vector of the incident radiation and the dipole moment of a vibration) and $D = d_{p,\parallel}/d_{p,\perp}$ ($d_{p,\parallel}$ is given by eq 3.48 and $d_{p,\perp}$ by eq 3.44). If α is taken as 0° for all bands that give the strongest intensity for p-polarized light and 90° for s-polarized light, the orientation function f will be 1 for all dipole moments perfectly aligned along the draw direction, −0.5 for transverse alignments, and 0 for random orientations. The orientation function can be correlated with the angle between the sample laboratory axis and the polymer chain axis through Herman's orientation function

$$f = (3\cos^2\theta - 1)/2 \tag{8.3}$$

Let us go back to the polyethylene data shown in Figure 8.27 and illustrate how dichroic ratios can be used to determine orientation of the crystalline phase near the surface. Although there are alternative transmission measurements using tilted angles (102) or polarization modulation experiments (103), ATR appears to be a straightforward approach, especially using p-polarized light (104). Figure 8.28 illustrates the dichroic ratios of the 730 cm^{-1} (A) and 720 cm^{-1} (B) bands in corrected ATR spectra plotted as a function of the penetration depth at 720 cm^{-1}. For comparison, the dichroic ratios of the 730 cm^{-1} (A') and 720 cm^{-1} (B') bands determined from transmission spectra are also shown.

To interpret the dichroic ratio results, it is necessary to know the relationship between the directions of the transition moment vectors of the IR bands of interest and the axes of the crystalline unit cell. For polyethylene, such data are available in the literature (105) and indicate that the direction

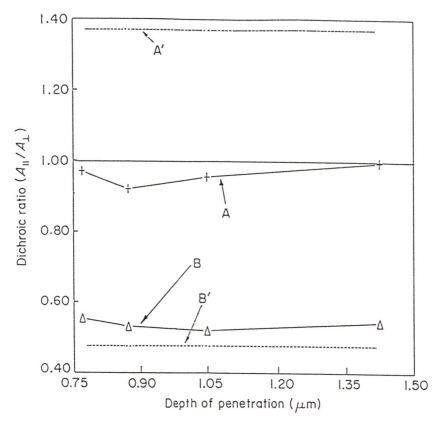

Figure 8.28. Dichroic ratios plotted as a function of depth of penetration: (A) 730 cm⁻¹ ATR band; (B) 720 cm⁻¹ ATR band, (A') 730 cm⁻¹ transmission band, and (B') 720 cm⁻¹ transmission band. (Reproduced with permission from reference 8. Copyright 1992 Butterworth–Heinemann, Ltd.)

of the transition moment vector of the 730 cm⁻¹ band coincides with the *a* axis of an orthorhombic unit cell for polyethylene, whereas for the crystalline portion of the 720 cm⁻¹ band the transition moment vector is aligned along the *b* axis. According to curve A in Figure 8.28, the dichroic ratio for 730 cm⁻¹ oscillates about unity. Because the transition moment vector for this band coincides with the *a* axis of the orthorhombic cell, these results indicate that the *a* axis is either oriented randomly or is preferentially perpendicular to the film plane direction. In contrast, curve A' shows that the dichroic ratio reaches values of 1.4, indicating that at the core of the film, to which transmission IR is sensitive, the *a* axis is predominantly oriented along the machine draw direction. Similarly, curves B and B' illustrate the dichroic ratio dependence for the band at 720 cm⁻¹ for ATR and transmission measurements, respectively. The fact that the dichroic ratios in all ATR and transmission spectra for the 720 cm^{-1} band are much below unity indicates that the *b* axis of the unit cell is predominantly oriented perpendicular to the machine draw direction.

As we have seen, the ATR band intensity should be corrected to reveal the effect of overlapping bands and intensity changes resulting from optical effects. Furthermore, to obtain orientation information, it is desirable to have X-ray data. The issue becomes more complicated when such data are not available, because when the sample is oriented, absorbance of a given band, or d_e, will depend not only on the number of species per volume or area but also on the structure of the species. Therefore, in addition to concentration changes, an orientation phenomenon becomes a dominating factor. As a matter of fact, the determination of orientation in polymer surfaces has been around for almost three decades. Flournoy (*105*) was the first to apply uniaxial stresses to isotactic polypropylene, and using parallel transmission and internal reflectance measurements, he concluded that the dichroic ratios in both experiments were equal, thus indicating the same orientations at the surface and in the bulk of the polymer.

In ATR FT-IR dichroism measurements, the polymer sample and the polarizer are rotated by 90°, allowing the detection of four spectra from which three absorbances (A_x, A_y, and A_z) are calculated:

$$A_{x,\perp} = \alpha A_x \qquad A_{z,\parallel} = \beta A_y + \gamma A_z$$
$$A_{y,\perp} = \alpha A_y \qquad A_{y,\parallel} = \beta A_y + \gamma A_z \tag{8.4}$$

where α, β, and γ are constants calculated from the knowledge of the sample and crystal refractive properties. They are given by (*103*)

$$\alpha = \frac{4n_{21}^2}{\tan\theta\left(1 - \dfrac{n_{21}^2}{\sin^2\theta}\right)^{1/2}(1 - n_{21})^2} \tag{8.5}$$

$$\beta = \frac{4n_{21}^2\left(1 - \dfrac{n_{21}^2}{\sin^2\theta}\right)}{\tan\theta\left(1 - \dfrac{n_{21}^2}{\sin^2\theta}\right)^{1/2}\left(1 - \dfrac{n_{21}^2}{\sin^2\theta} + n_{21}^4\cot^2\theta\right)} \tag{8.6}$$

$$\gamma = \frac{4n_{21}^2}{\tan\theta\left(1 - \dfrac{n_{21}^2}{\sin^2\theta}\right)^{1/2}\left(1 - \dfrac{n_{21}^2}{\sin^2\theta} + n_{21}^4\cot^2\theta\right)} \tag{8.7}$$

where $n_{21} = n_2/n_1$.

It is well established that IR band intensity as a function of wavenumber can be calculated from perturbation theory. In this case it is necessary to consider that the intensity is proportional to the square of the scalar product of the transition dipole moment $\mathbf{M} = \mu/q$ and the electric vector $\mathbf{E}$ of radiation, or $(\mathbf{M} \cdot \mathbf{E})^2$.

The dipole moment $\mathbf{M}$ is the change of the molecular dipole moment during the vibrational transition, and q is the corresponding vibrational normal coordinate. Because the transition dipole moment $\mathbf{M}$ and the electric field have certain spatial positions, they can be expressed in x, y, and z space:

$$(\mathbf{M} \cdot \mathbf{E})^2 = (M_x E_x + M_y E_y + M_z E_z)^2 \tag{8.8}$$

and this relationship represents the band area. However, to perform structural analysis, it is necessary to know the orientation of $\mathbf{M}$ with respect to molecular coordinates (a, b, c), and for that reason the relative orientation of the x–y–z and a–b–c coordinates and the orientation of the transition dipole moment $\mathbf{M}$ with respect to a, b, and c must be known. This last condition—that is, the transformation of M_x, M_y, M_z into M_a, M_b, M_c—often requires either an experiment or normal coordinate calculations (*106*). Once M_x, M_y, and M_z are calculated, they are inserted into eq 8.8.

A partial axial orientation of a molecule with a dipole moment $\mathbf{M}$ and a, b, and c axes is depicted in Figure 8.29. The angles ϕ, θ, and γ are referred to as the Eulerian angles, which allow a precise description of the position of two coordinate systems with respect to each other. Depending on these angles, various structural entities can be formed. For example, when the angles ϕ, θ, and γ are fixed, an oriented single crystalline ultrastructure exists; when ϕ is isotropic from $0°$ to $360°$, and θ and γ are fixed, a random arrangement of a large number of microcrystallites around the z axis exists; when ϕ and θ are isotropic and γ varies from $0°$ to $90°$, a random arrangement of molecules around the z axis and free rotation around a molecular c axis exist; when ϕ, θ, and γ are isotropic, random orientation is expected.

Each of these ultrastructures may be represented by a unique distribution function. We need to look for the probability of finding the molecular axis in space that would correspond to the axis of a given vibration. The two most common distribution functions are the delta and Kratky distribution functions (*107–109*). The delta distribution function is given by

$$f(\gamma) = \delta(\gamma - \gamma_0) \tag{8.9}$$

This function represents a cone shape, if all axes form a fixed angle γ_0 with the z axis and are isotropically distributed around the z axis.

The Kratky distribution function is a more complex function

$$f(\gamma) = v^{3/4} \sin \gamma \, / \, (v^{-3/2} \cos^2\gamma + v^{3/2} \sin^2\gamma)^{3/2} \tag{8.10}$$

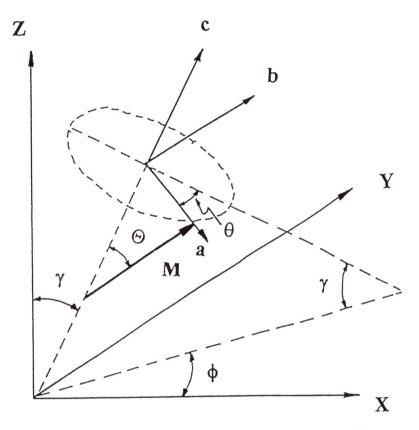

*Figure 8.29. Orientation of molecular (abc) coordinates and Cartesian (xyz) coordinates. For accurate structural analysis, a position of the transition dipole moment **M** with respect to abc and xyz coordinates and relative positions of the coordinates should be known.*

and incorporates the molecular polymer's ordering upon stretching, which includes a draw ratio $L = L/L'$ (not to be confused with frequency ν) where L is the film length after stretching and L' is the length before. Further discussions concerning these functions can be found in the literature (*110*).

The surface orientation studies conducted by Mirabella (*13, 14*) on uniaxially extended polypropylene films showed that the use of the raw absorbencies of the 841 and 809 cm^{-1} bands can be beneficial. Unlike other studies, the ratio of two bands from the same measurement was taken in estimating the orientation function, giving an advantage of maintaining the same contact between the sample and the crystal. This issue was resolved earlier by Sung (*111*), who developed a rotatable ATR cell in which both measurements can be made without demounting and removing the sample. Furthermore, Sung et al. (*112*) also identified two IR bands at 998 and 973 cm^{-1}, which represent crystalline and average orientations and can be used as a measure of how the orientation function changes with the draw ratio of drawn polypropylene.

As was illustrated earlier, such measurements should involve detection of polarized ATR at many horizontal angles. For that reason, Yuan and Sung (*113*) proposed an elegant approach that was a truncated hemispherical crystal placed on a goniometer, allowing complete horizontal rotation. Figure 8.30 is a schematic diagram of this attachment, which was used to determine orientation changes with drawing in poly(ethylene terephthalate) (*114*).

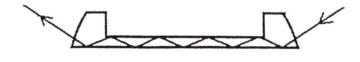

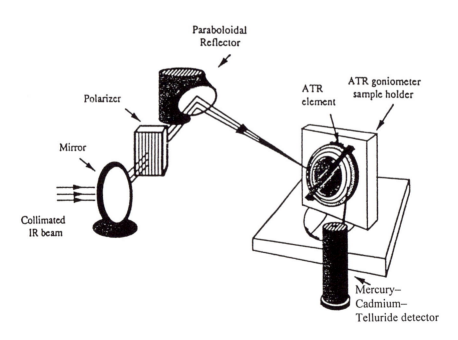

Figure 8.30. A schematic diagram of optical arrangement showing an ATR crystal and its holder in a FT-IR instrument. (Reproduced with permission from reference 111. Copyright 1991 American Chemical Society.)

References

1. Nikitin, V. N.; Pokrovskii, E. I. *Dokl. Akad. Nauk SSSR* **1954**, *95*, 109.

2. Tobin, M. C.; Carrano, M. J. *J. Polym. Sci.* **1957**, *24*, 93.

3. Wedgewood, A. R.; Seferis, J. C. *Pure Appl. Chem.* **1983**, *55*, 873.

4. Okada, T.; Mandelkern, L. *J. Polym. Sci. Polym. Phys. Ed.* **1967**, *5*, 239.

5. Zerbi, G.; Gallino, G.; Fanti, N. D.; Baini, L. *Polymer* **1989**, *30*, 2324.

6. Dothee, D.; Vigourreux, J. M.; Camelot, M. *Polym. Degrad. Stab.* **1988**, *22*, 161.

7. Abbate, S.; Gussoni, M.; Zerbi, G. *J. Chem. Phys.* **1979**, *70*, 3577.

8. Huang, J. B.; Hong, J. W.; Urban, M. W. *Polymer* **1992**, *33*, 5173.

9. Hawranek, J. P.; Jones, R. N. *Spectrochim. Acta* **1976**, *32A*, 99.

10. Graf, R. T.; Koenig, J. L.; Ishida, H. In *Fourier Transform Infrared Characterization of Polymers*; Ishida, H., Ed.; Plenum: New York, 1987.

11. Bardwell, J. A.; Dignam, M. J. *Anal. Chim. Acta* **1986**, *181*, 253.

12. Bertie, J. E.; Eysel, H. H. *Appl. Spectrosc.* **1985**, *39*, 392.

13. Mirabella, F. M. *J. Polym. Sci. Polym. Phys. Ed.* **1984**, *22*, 1283.

14. Mirabella, F. M. *J. Polym. Sci. Polym. Phys. Ed.* **1984**, *22*, 1293.

15. Nishio, E.; Morimoto, M.; Nishikida, K. *Appl. Spectrosc.* **1990**, *44*, 1639.

16. Walls, D. J. *Appl. Spectrosc.* **1991**, *45*, 1193.

17. Wentink, T.; Willwerth, L. J.; Phaneuf, J. P. *J. Polym. Sci.* **1961**, *55*, 551.

18. Lovinger, A. J.; Freed, D. J. *Macromolecules* **1980**, *13*, 989.

19. Hahn, B.; Percec, V. *J. Polym. Sci. Polm. Chem. Ed.* **1987**, *25*, 781.

20. Urban, M. W.; Salazar-Rojas, E. M. *Macromolecules* **1988**, *21*, 372.

21. Kuhn, K. J.; Hahn, B.; Percec, V.; Urban, M. W. *Appl. Spectrosc.* **1987**, *41*, 843.

22. Exsted, B. J.; Urban, M. W. *J. Inorg. Organomet. Polym.* **1993**, *3*, 105.

23. Schmitt, B. R.; Kim, H.; Urban, M. W. *Langmuir* **1996**, in press.

24. Coleman, M. M.; Painter, P. C. *J. Macromol. Sci. Rev. Macromol. Mater. Res.* **1977**, *C16*, 197.

25. Jacobsen, R. J. In *Fourier Transform Infrared Spectroscopy: Applications to Chemical Systems*; Ferraro, J. R.; Basile, L. J., Eds.; Academic: New York, 1979; and references therein.

26. Hirschfeld, T. *Appl. Spectrosc.* **1977**, *31*, 289.

27. Saucy, A.; Simko, S. J.; Linton, R. W. *Anal. Chem.* **1985**, *57*, 871.

28. Hirayama, T.; Urban, M. W. *Prog. Org. Coat.* **1992**, *20*, 81.

29. He, M.; Urban, M. W.; Bauer, R. *J. Appl. Polym. Sci.* **1993**, *49*, 345.

30. Mittlefehldt, E. R.; Gardella, A., Jr. *Appl. Spectrosc.* **1989**, *43*, 1172.

31. Kauppinen, J. K.; Moffatt, D. J.; Mantsch, H. H.; Cameron, D. G. *Appl. Spectrosc.* **1981**, *35*, 271.

32. Triolo, P. M.; Andrade, J. D. *J. Biomed. Mater. Res.* **1983**, *17*, 129.

33. Hoffman, A. S.; Horbett, T. A.; Ratner, B. D.; Hanson, S. R.; Karker, L. A.; Reynolds, L. O. In *Biomaterials: Interfacial Phenomena and Applications*; Coopes, S. L.; Peppas, N. A., Eds.; American Chemical Society: Washington, DC, 1982.

34. Becker, G. E.; Gobeli, G. W. *J. Chem. Phys.* **1963**, *38*, 2942.

35. Beckmann, K. H.; Harrick, N. J. *J. Electrochem. Soc.* **1971**, *118*, 614.

36. Chabal, Y. J. In *Internal Reflection Spectroscopy*; Mirabella, F. M., Ed.; Dekker: New York, 1993.

37. Jakob, P.; Dumas, P.; Chabal, Y. J. *J. Appl. Phys. Lett.* **1991**, *59*, 2968.

38. Jakob, P.; Chabal, Y. J. *J. Chem. Phys.* **1991**, *95*, 2897.

39. Reutt-Robey, J. E.; Doren, D. J.; Chabal, Y. J.; Christman, S. B. *Phys. Rev. Lett.* **1988**, *61*, 2778.

40. Reutt-Robey, J. E.; Doren, D. J.; Chabal, Y. J.; Christman, S. B. *J. Chem. Phys.* **1990**, *93*, 9113.

41. Swanson, J. R.; Friend, C. M.; Chabal, Y. J. *J. Chem. Phys.* **1987**, *87*, 5028.

42. Gaboury, S. R.; Urban, M. W. *Polym. Comm.* **1991**, *32*, 390.

43. Gaboury, S. R.; Urban, M. W. In *Structure–Property Relations in Polymers: Spectroscopy and Performance*. Advances in Chemistry Series 236; Urban, M. W.; Craver, C. D., Eds.; American Chemical Society: Washington, DC, 1993; Chapter 34.

44. Gaboury, S. R.; Urban, M. W. *Langmuir* **1993**, *9*, 3225.

45. Gaboury, S. R.; Urban, M. W. *Langmuir* **1994**, *10*, 2289.

46. Kim, H.; Urban, M. W. *Langmuir* **1995**, *11*, 2071.

47. Kim, H.; Urban, M. W. *Langmuir* **1996**, *12*, 1047.

48. Thorstenson, T. A.; Tebelius, L. K.; Urban, M. W. *J. Appl. Polym. Sci.* **1993**, *50*, 1207.

49. Evanson, K. W.; Urban, M. W. In *Surface Phenomena and Fine Particles in Water-Based Coatings and Printing Technology*; Sharma, M. K.; Micale, F. J., Eds.; Plenum: New York, 1991.

50. Thorstenson, T. A.; Evanson, K. W.; Urban, M. W. *Structure–Property Relations in Polymers: Spectroscopy and Performance.* Advances in Chemistry Series 236; Urban, M. W.; Craver, C. D., Eds.; American Chemical Society: Washington, DC, 1993; Chapter 11.

51. Urban, M. W.; Evanson, K. W. *Polym. Comm.* **1990**, *31*, 279.

52. Evanson, K. W.; Urban, M. W. *J. Appl. Polym. Sci.* **1991**, *42*, 2287.

53. Evanson, K. W.; Urban, M. W. *J. Appl. Polym. Sci.* **1991**, *42*, 2297.

54. Evanson, K. W.; Urban, M. W. *J. Appl. Polym. Sci.* **1991**, *42*, 2309.

55. Thorstenson, T. A.; Urban, M. W. *J. Appl. Polym. Sci.* **1993**, *47*, 1387.

56. Thorstenson, T. A.; Tebelius, L. K.; Urban, M. W. *J. Appl. Polym. Sci.* **1993**, *49*, 103.

57. Kunkel, J. P.; Urban, M. W. *J. Appl. Polym. Sci.* **1993**, *50*, 1217.

58. Fina, L. J.; Chen, G.; Valentini, J. E. *Ind. Eng. Chem. Res.* **1992**, *31*, 1659.

59. Leidheiser, H.; Funke, W. *J. Oil Colour Chem. Assoc.* **1992**, *70*, 121.

60. Leidheiser, H., Jr. *Corrosion* **1983**, *39*, 189.

61. Mattson, J. S.; Mark, H. B. *Activated Carbon*; Dekker: New York, 1971.

62. Stevenson, W. T. K.; Garton, A. *J. Mater. Sci. Lett.* **1987**, *6*, 580.

63. Ishino, Y.; Ishida, H. *Anal. Chem.* **1986**, *58*, 2448.

64. Ishino, Y.; Ishida, H. *Appl. Spectrosc.* **1988**, *42*, 1296.

65. Sellitti, C.; Koenig, J. L.; Ishida, H. *Appl. Spectrosc.* **1990**, *44*, 830.

66. Hart, W. W.; Painter, P. C.; Koenig, J. L.; Coleman, M. M. *Appl. Spectrosc.* **1977**, *31*, 220.

67. Rositani, F.; Antonucci, P. L.; Minutoli, M.; Giordano, N. *Carbon* **1981**, *25*, 325.

68. Claybourn, M.; Colombel, P.; Chalmers, J. *Appl. Spectrosc.* **1991**, *45*, 279.

69. Garton, A. *J. Polm. Sci. Polym. Chem. Ed.* **1985**, *22*, 1495.

70. Garton, A. *Polym. Compos.* **1984**, *5*, 258.

71. Garton, A.; Daly, J. H. *J. Polym. Sci. Polym. Chem. Ed.* **1985**, *23*, 1031.

72. Tiefenthaler, A. M.; Urban, M. W. *Appl. Spectrosc.* **1988**, *42*, 163.

73. He, M.; Urban, M. W.; Bauer, R. S. *J. Appl. Polym. Sci.* **1993**, *49*, 345.

74. Brandrup, J.; Immergut, E. H. *Polymer Handbook*, 3rd ed.; Wiley: New York, 1989.

75. Chan, M. G.; Allara, D. L. *Polym. Eng. Sci.* **1974**, *4*, 12.

76. Chan, M. G.; Allara, D. L. *J. Colloid Interface Sci.* **1974**, *47*, 697.

77. Cullis, C. F.; Berr, H. S. *Eur. Polym. J.* **1978**, *14*, 575.

78. Dickie, R. A.; Hammond, J. S.; Holubka, J. W. *Ind. Eng. Chem. Prod. Res. Dev.* **1981**, *20*, 339.

79. Dickie, R. A.; Carter, R. O.; Hammond, J. S.; Parsons, J. L.; Holubka, J. W. *Ind. Eng. Chem. Prod. Res. Dev.* **1984**, *23*, 297.

80. Urban, M. W. *Vibrational Spectroscopy of Molecules and Macromolecules on Surfaces*; Wiley: New York, 1993.

81. Nguyen, T.; Byrd, E.; Lin, C. *J. Adhes. Sci. Technol.* **1991**, *5*, 697.

82. Van Oeteren, K. A. *Werkst. Korros.* **1964**, *10*, 427.

83. Makishima, H.; Toyoda, T.; Nakamula, N. *Shikazai Kyokaishi* **1971**, *44*, 156.

84. O'Brien, R. N.; Hartman, K. *Pap. Meet. Am. Chem. Soc. Div. Org. Coat. Plast. Chem.* **1968**, *28*, 236.

85. Urban, M. W.; Stewart, M. T. *J. Appl. Polym. Sci.* **1990**, *39*, 265.

86. Stewart, M. T.; Urban, M. W. *Polym. Mater. Sci. Eng.* **1988**, *59*, 334.

87. Inagaki, N.; Oh-Ishi, K. *J. Polym. Sci. Polym. Chem. Ed.* **1985**, *23*, 1445.

88. Rimai, D. S.; DeMejo, L. P.; Vreeland, W.; Bowen, R.; Gaboury, S. R.; Urban, M. W. *J. Appl. Phys.* **1992**, *71*, 2253.

89. Gaboury, S. R.; Urban, M. W. *J. Appl. Polym. Sci.* **1992**, *44*, 401.

90. Pesetskii, S. S.; Egorenkov, N. L.; Shcherbakov, S. V. *Kolloidn. Zh.* **1981**, *43*, 992.

91. Ahearn, J. S.; Davis, G. D.; Sun, T. S.; Venables, J. D. In *Adhesion Aspects of Polymeric Coatings*; Mittal, K. L., Ed.; Plenum: New York, 1983.

92. Hennemann, O. D.; Brockmann, W. *J. Adhes.* **1981**, *12*, 297.

93. Davis, G. D.; Sun, T. S.; Ahearn, J. S.; Venables, J. D. *J. Mater. Sci.* **1982**, *17*, 1807.

94. Golander, C. G.; Sultan, B. A. *J. Adhes. Sci. Technol.* **1988**, *2*, 125.

95. Erickson, J. C.; Golander, C. G.; Baszkin, A.; Ter-Minassian-Saraga, L. *J. Colloid Interface Sci.* **1983**, *100*, 381.

96. Urban, M. W.; Koenig, J. L. *Appl. Spectrosc.* **1987**, *41*, 1028.

97. Colthup, N. B.; Daly, L. H.; Wiberley, S. E. *Introduction to Infrared and Raman Spectroscopy*; Academic: London, 1975.

98. Mittal, K. L., Ed. *Adhesion Aspects of Polymeric Coatings*; Plenum: New York, 1983; and references therein.

99. Lynch, B. M.; Chisholm, S. L. *Langmuir* **1992**, *8*, 351.

100. Crawford, B. In *Advances in Infrared and Raman Spectroscopy*; Clark, R. J. H.; Hester, R. E., Eds.; Heyden: London, 1978; Vol. 4.

101. Bardwell, J. A.; Dignam, M. *J. Anal. Chim. Acta* **1988**, *181*, 253.

102. Fina, L. J.; Koenig, J. L. *J. Polym. Sci. Polym. Phys. Ed.* **1986**, *24*, 2509.

103. Stein, R. S. *J. Appl. Polym. Sci.* **1961**, *5*, 96.

104. Hobbs, J. P.; Sung, C. S. P.; Krishnan, K.; Hill, S. *Macromolecules* **1983**, *16*, 193.

105. Flournoy, P. A. *Spectrochim. Acta* **1966**, *22*, 15.

106. Wilson, E. B.; Decius, J. C.; Cross, P. C. *Molecular Vibrations: The Theory of Infrared and Raman Vibrational Spectra*; McGraw–Hill: New York, 1965.

107. Zbinden, R. *Infrared Spectroscopy of High Solids*; Academic: New York, 1964.

108. Kratky, O. *Kolloid Z.* **1933**, *64*, 213.

109. Michl, J.; Thulstrup, E. W. *Spectroscopy with Polarized Light*; VCH: New York, 1986.

110. Fringeli, U. P. In *Internal Reflection Spectroscopy*; Mirabella, F. M., Jr., Ed.; Dekker: New York, 1993.

111. Sung, C. S. P. *Macromolecules* **1981**, *14*, 591.

112. Sung, N. H.; Lee, H. Y.; Yuan, P.; Sung, C. S. P. *Polym. Eng. Sci.* **1989**, *29*, 791.

113. Yuan, P.; Sung, C. S. P. *Macromolecules* **1991**, *24*, 6095.

114. Lee, K. H.; Sung, C. S. P. *Macromolecules* **1993**, *26*, 3289.

9

In Situ ATR Experiments

ATR Polymerization Kinetics

During the polymerization process, molecules react to form larger segments, and as a result the nonbonding distances between monomers prior to the reaction become bonding distances between the repeating units which are smaller. This is reflected in changes in the density, molar refraction, and refractive index. For example, when methyl methacrylate polymerizes, the density change $\Delta\rho/\rho$ is about 26%. For styrene, the change is about 17%. Although it is easily visualized that, as a result of going from nonbonding to bonding distances, volume changes occur, it is often forgotten that when such polymerization, or cross-linking, reactions occur on the surface they result in shrinkage. Because substrates usually do not change their volume, tremendous interfacial stresses may be induced. In a polymerization process on a surface of an ATR crystal, a similar scenario may occur, and the spectroscopic changes may result not only from the polymerization but also from stresses induced by shrinkage. Therefore, it is a good practice to conduct parallel experiments in transmission mode in order to eliminate such possibilities.

One of the first in situ kinetic polymerization studies on nylon-6 using reaction injection molding (RIM) was reported by Ishida and Scott (1). Figure 9.1 illustrates the three-dimensional representation of the amide I and II bands at 1660 and 1540 cm^{-1} of the monomeric precursor ε-caprolactam and the formed nylon-6. Whereas the enhanced band intensity at the later stages of the polymerization results from the thermally induced crystallization, a drastic increase of the amide II band indicates a fast polymerization process. Using this approach, it is possible to obtain in situ molecular information on polymerization, crystallization, and chain packing, thus avoiding many difficulties encountered while using thermal methods. Another example is in situ cross-linking of photoresists (2). The experimental setup shown in Figure 9.2 was used to examine curing processes of a polyester–photoresist sandwich attached to an ATR element. Again, this is an example

3348–9/96/0185/$15.00/0/© 1996 American Chemical Society

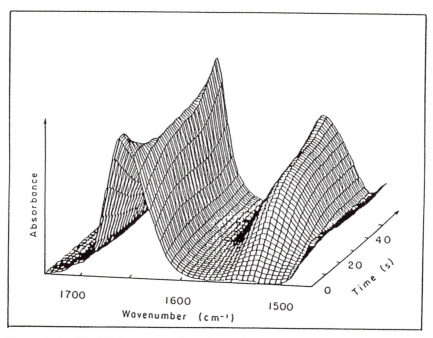

Figure 9.1. ATR FT-IR spectra plotted as a function of time during the reaction injection molding of nylon-6. (Reproduced with permission from reference 1. Copyright 1987 Rubber Chemistry and Technology.)

of a useful experimental setup that cannot be used for quantitative purposes because of the spectral distortions resulting from the sample–crystal contact.

ATR Flow-Through Cells

Liquids can be successfully analyzed with a circular cell. This cell is only one example among many, such as those illustrated in Figure 9.3, that can be used in continuous flow situations. Even high-pressure experiments—for example, on hydraulic fluids—can be done with such attachments. One issue that the potential user should be aware of is a seal between the crystal and the cell cavity. Such a seal, usually in a form of an O-ring, also touches the crystal and thus absorbs the incident radiation. Although one could relatively safely assume that the surface area in contact with the crystal is small, and that no absorptions due to the O-ring material should be detected, in studies in which the detection limits are being approached, the presence of an O-ring may be of great significance. The most common

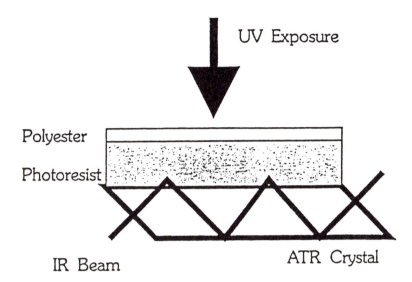

UV Exposure

Polyester

Photoresist

IR Beam ATR Crystal

Figure 9.2. Schematic diagram of an ATR crystal and photoresist sandwich. UV exposure is permitted through the polyester film side of the melamine. (Reproduced with permission from reference 2. Copyright 1992 Society for Applied Spectroscopy.)

material used for O-rings is poly(tetrafluoroethylene), since it provides good seal and has no bands in the C–H stretching region. The only band that may obscure the spectral measurements is the one around 1250 cm^{-1} due to the C–F stretching modes. One of the important experimental considerations in ATR measurements is the contact between the sample and the ATR crystal. Although the most common practice is to use a rectangular crystal, several applications with circular crystals have been demonstrated. A circular cell, schematically depicted in Figure 9.3C, originally developed to study aqueous solutions (*3, 4*) and used for polymeric systems (*5, 6*), can also be used for studies of films and fibers (*7*). A cylindrical ATR cell can be used to analyze diluted antibiotics (*8*), natural products (*9*), and silane coupling agents (*10*).

ATR Immersion Probes and Biological Applications

Developments in fiber optics technology and the ability to immerse ATR elements in reaction vessels opened unlimited opportunities for in situ studies as well as quality control in pilot plants. Among the most illustrative examples is the deep immersion probe sensing head, which is schematically illustrated in Figure 9.4A (b), along with other approaches outlined in

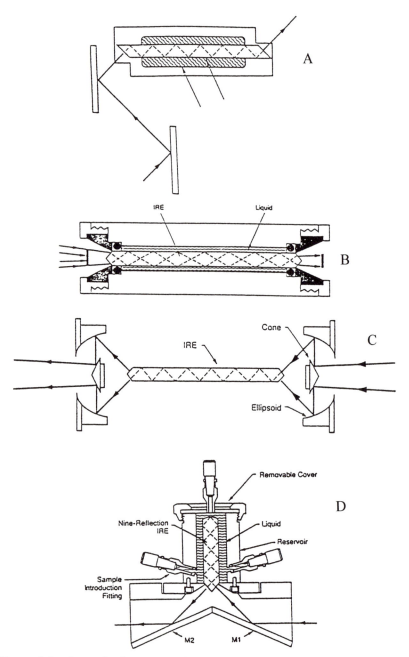

Figure 9.3. Optical schematics of several commercially available ATR cells: (A) Squarecol (Specac), (B) Tunnel (Axion), (C) Circle (SpectraTech), and (D) Liquid Prism (Harrick).

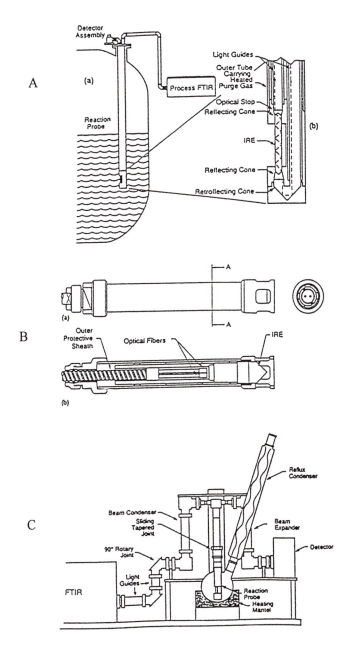

Figure 9.4. Examples of in situ ATR accessories: (A) Axion ATR reaction probe (b) with a reaction vessel (a); (B) external view (a) and optical diagram of a fiber optics ATR (Specac) dipping probe; and (C) schematic diagram of the Axion laboratory-scale reaction probe. (Reproduced with permission from reference 11. Copyright 1993 Marcel Dekker.)

the figure caption. In essence, each design consists of a circular ATR element that, with the proper geometry, may serve as a sensor in numerous applications.

In many biological applications, water is used as a solvent. One of the drawbacks of IR spectroscopy is the fact that the vibrating atoms in an H_2O molecule undergo extremely large changes in dipole moment during their vibrations. As a result, when water is present, excessively strong absorption bands near 3500 cm^{-1} (O–H stretching) and 1640 cm^{-1} (O–H bending) are detected in an IR spectrum. As an example, Figure 9.5 illustrates the refractive and absorption index spectra of H_2O in the O–H stretching and bending regions. In this case, the spectra were recorded using a circular ATR element because in such an experimental arrangement, the water bands are weaker, thus giving higher intensities of other species present in an aqueous solution.

For biological molecules, the environmental conditions, such as pH, ionic strength, and other solution properties, may influence conformational and structural changes. It is not only the difficulty arising from the complexity of a repeating unit but also other properties (for example, adsorption) that may be of a prime significance in ATR measurements. If a given protein is adsorbed on an ATR crystal during the spectral collection, the spectrum of the adsorbed molecule will be very different from the spectrum of the same molecule in a solution. In this case, what often occurs is band broadening. Some proteins are mixtures of various species, which may be irreversibly or reversibly adsorbed on an ATR crystal. In this case, spectral deconvolution methods should be useful, but only after an ATR spectrum has been properly converted to its adsorption counterpart. It is always desirable to conduct experiments at several temperatures and concentrations, as these variables will affect the amount of a species adsorbed on a surface. Tickanen et al. (*12*) examined how interactions between the internal reflection element and particles influence absorption values. The study demonstrated that the ATR band intensities of aqueous suspensions may be affected by such interactions. Apparently, electrostatic charges on the particle surfaces and the ATR crystal play a key role in a quantitative analysis.

Adsorption of human serum albumin onto contact lenses or artificial organs can also be examined by ATR. The method can detect not only transient effects in the adsorption process but also conformational changes resulting from adsorption at various time intervals. As was demonstrated by Castillo et al. (*13*), hydrogen bonding and the involvement of the hydrophobic side are the primary sources of adsorption. In contrast, other studies indicated that albumin and prothrombin adsorbed on silica retain their native conformation, but the conformation of γ-globulin is altered. Several earlier ATR studies claimed that albumin retains its native conformation upon absorption (*14*). It should be kept in mind that the earlier experi-

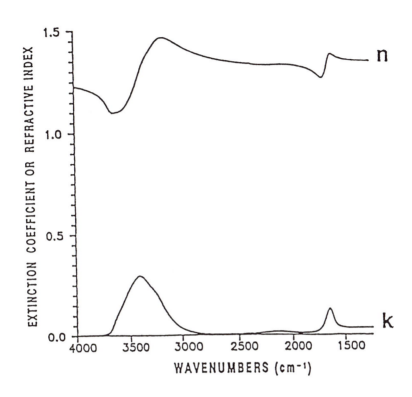

Figure 9.5. Absorption and refractive index spectra of water.

ments did not employ FT-IR instruments, so sensitivity was an issue. A series of studies revealing many structural details conducted by Gendreau and co-workers (*15–17*) used ATR FT-IR, and detected albumin denaturation with increased adsorption time, along with conformational changes. ATR FT-IR can be a highly useful source of information in the analysis of agricultural products (*18*) and synthetic polymers (*5*) suspended in aqueous solutions. Discussion of the vibrational features of water-soluble polymers is available in the literature (*19*).

In an effort to enhance the sensitivity of the optical designs of probes with circular ATR elements, *filtrometers* were proposed (*20*). An optical diagram of such an on-line system is illustrated in Figure 9.6. Because the collecting mirrors are *f*/0.6, a high energy throughput can be achieved, and the optical efficiency of focusing the energy into the crystal is better than 50%. As a result, the signal-to-noise ratio is tremendously improved. Such a configuration is especially useful in monitoring components of aqueous streams, such as dissolved sugar or carbon dioxide in beverages, or acid group content in fruit juices. Similar applications can be envisioned in

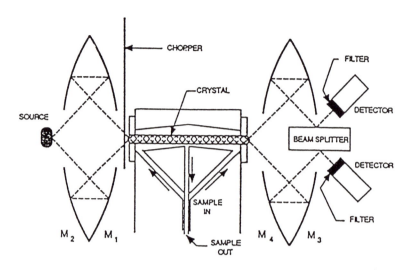

Figure 9.6. Schematic diagram leading to enhancement of sensitivity in a flow cell. (Reproduced with permission from reference 11. Copyright 1993 Marcel Dekker.)

pharmaceutical and nuclear chemistry, where mixing processes or heavy water concentrations need to be determined. The primary limitation, however, is spectral resolution; the method is suitable only for well-isolated, broad bands.

Electrochemical Processes

Although many attempts were made to use internal reflection spectroscopy to monitor electrochemical processes, they were not successful until the optically transparent electrode, often called optically transparent thin-layer electrode (OTTLE), was embedded between the internal reflection crystal and the sample layer. Figure 9.7 is a schematic diagram of the electrochemical cell. As the voltage is applied, it initiates an electrochemical reaction at the thin film electrode surface, and spectroscopic changes are monitored. There are numerous reports in the literature describing various OTTLE structures (*21–25*). Although numerous cell designs and configurations provide high sensitivity (*26*), the concept of having the reference beam parallel to the sample beam and penetrating the crystal had proved to be a good one (*27*). In this approach, the cell compartment is half filled and the sample beam strikes a lower filled part of the cell. At the same time, the upper beam penetrates part of the crystal, thus providing a suitable reference for the sample spectrum.

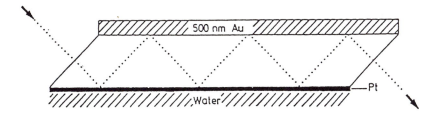

Figure 9.7. Internal reflection element with a metallic mirror used in electrochemical cells: 500 nm gold serves as a mirror; platinum layer is an optically transparent film. (Reproduced with permission from reference 11. Copyright 1993 Marcel Dekker.)

   To study reactions at electrochemical phase boundaries, Muller and Abraham-Fuchs (*28*) suggested the experimental setup shown in Figure 9.8. In this arrangement, the ATR crystal (1) is coated with a thin metallic film (2) and electrolyte (3). The film is connected through the circuit to an electrode immersed in the electrolyte solution. One requirement is that the depth of penetration of the electromagnetic radiation must be greater than the metallic film thickness. Under such conditions the radiation will interact with the metal electrode boundary layer. The metallic layer is around 50 to 500 Å thick.

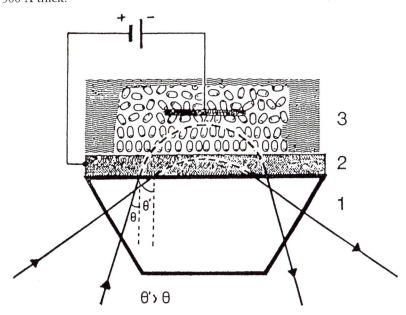

Figure 9.8. An experimental setup to study electrochemical phase boundaries. (Reproduced with permission from reference 11. Copyright 1993 Marcel Dekker.)

The chemical reactions at solid–liquid interfaces are particularly important in electrochemical processes. ATR has been used to study oxidation of poly(thiophene) and poly(3-methylthiophene) on Pt (*29*), adsorption of SCN⁻ from aqueous solutions on Pt (*30*), solvent–electrolyte interaction for acetonitrile solutions on Pt (*31*), electrode processes at semiconductor–solution interfaces (*32*), adsorption of proteins under flow on an ATR crystal (*33*, *34*), adsorption of protein on silicone (*35*), orientation of cadmium arachidate on a glass plate (*36*), and adsorption and oxidation of $Fe(CN)_6^{4-}$ ions on Pt (*37*).

As discussed in earlier chapters, a key feature of an ATR experiment is the refractive index ratio between the sample and an ATR element. That ratio (or difference) should be such that the refractive index of the sample should be smaller, so an evanescent wave at the ATR element and the sample can be formed. The same principle applies to light propagating through fiber optics. Simohony et al. (*38*) used this approach and showed that an experimental setup like that in Figure 9.9 can be used to detect as little as 6 ng (or approximately 1/20 of a monolayer) of a known species, and a high-quality spectrum can be recorded with as little as 10 µg.

Langmuir–Blodgett Films

Under appropriate conditions, water-insoluble amphiphilic molecules spread at the air–water interface can be transferred to a solid support, leading to a formation of Langmuir–Blodgett films. Continuously growing interest in preparation and characterization of Langmuir–Blodgett films (*39*, *40*) stimulated intense studies in this area. Although application studies concerning electron conduction, biomaterials, or semiconductivity were of the primary interest (*41–43*), numerous methods and analytical approaches, including X-ray photoelectron spectroscopy (*32*), X-ray diffraction, and photoacoustic FT-IR (*44*) were used. For obvious reasons, however, ATR did not receive much attention. One interesting application was reported by Miyano et al. (*45*), who used an internal reflection accessory to examine monolayer structures of dioctadecyldimethylammonium bromide right above a water surface. Figure 9.10 shows an experimental setup for determining adsorption kinetics.

Diffusion

A common approach to measuring diffusion of small molecules in polymers is to place a polymer film of known dimensions in a penetrant bath that can be considered infinite in comparison with the polymer thickness and let the penetrant sorb into the polymer network. By measuring the

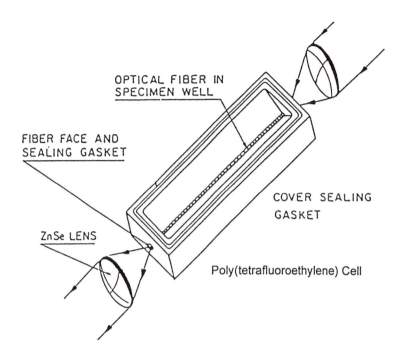

OPTICAL FIBER IN
SPECIMEN WELL

FIBER FACE AND
SEALING GASKET

COVER SEALING
GASKET

ZnSe LENS

Poly(tetrafluoroethylene) Cell

Figure 9.9. Schematic diagram of silver halide fiber optics used for analysis of organic solutions and solids. (Reproduced with permission from reference 37. Copyright 1988 American Chemical Society.)

weight loss or gain, or concentration changes as a function of time, and applying sorption kinetics, we can determine diffusion properties.

Let us assume that the polymer thickness is $2L$. When the specimen is placed into a penetrant bath, the concentration C_s at two surfaces $2L$ apart from each other will be immediately established. If the initial concentration of the penetrant is equal to zero, at time t at any distance from each surface a new concentration C will be established (46) as given by

$$\frac{C}{C_s} = 1 - \frac{4}{\pi}\frac{(-1)^n}{2n+1}\exp\left[\frac{-D(2n+1)^2\pi^2 t}{4L^2}\right]\cos\left[\frac{(2n+1)\pi z}{2L}\right] \quad (9.1)$$

By integrating eq 9.1 over the thickness of the film, we obtain the sorbed mass as

$$\frac{M_t}{M_\infty} = 1 - \sum_{n=0}^{\infty}\frac{8}{(2n+1)^2\pi^2}\exp\left[\frac{-D(2n+1)^2\pi^2 t}{4L^2}\right] \quad (9.2)$$

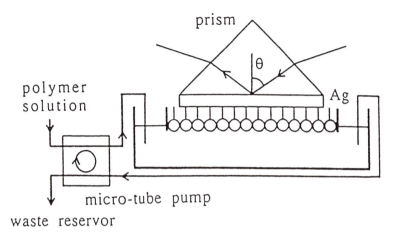

Figure 9.10. In situ measurements of Langmuir–Blodgett films by internal reflection approach.

where M_t is the mass sorbed at time t, and M_∞ is the mass sorbed at equilibrium. For short penetrant exposure times, this equation is usually simplified to

$$\frac{M_t}{M_\infty} = \frac{2}{L}\left(\frac{D}{\pi}\right)^{1/2} t^{1/2} \tag{9.3}$$

If M_t/M_∞ is plotted as a function of $t^{1/2}$, a diffusion coefficient can be determined from the slope, providing the plot is linear.

Although these considerations lead to fairly well understood sorption kinetics, experimental approaches are difficult and therefore not very accurate or reliable. For that reason Fieldson and Barbari (47) proposed another approach, which allows in situ ATR measurement of diffusion of small molecules in thin polymer films. The sampling configuration they proposed is illustrated in Figure 9.11. In their experiment, only one surface of the film was exposed to a penetrating fluid, and the other side of the crystal was used to control temperature. Fieldson and Barbari also proposed combining the Fickian concentration profile (eqs 9.1 and 9.2) with the ATR band intensity equation derived for weak bands. By simplifying the results, they obtained

$$\ln\left(1 - \frac{A_t}{A_\infty}\right) = \ln\left(\frac{4}{\pi}\right) - \frac{D\pi^2}{4L^2 t} \tag{9.4}$$

where A_t and A_∞ are the band intensities analogous to C_t and C_∞ or M_t and M_∞. This equation allows us to determine the diffusion coefficient by

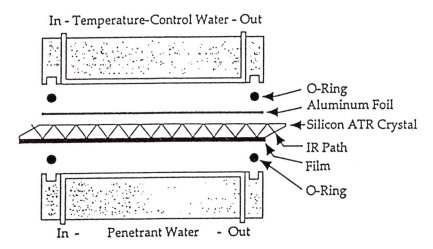

In - Temperature-Control Water - Out

O-Ring
Aluminum Foil
Silicon ATR Crystal
IR Path
Film
O-Ring

In - Penetrant Water - Out

Figure 9.11. ATR cell design for monitoring diffusion through polymer films. (Reproduced with permission from reference 11. Copyright 1993 Butterworth–Heinemann, Ltd.)

plotting the logarithm of the ATR intensities as a function of time. Note that several approximations were made in deriving this relationship. It was assumed that the depth of penetration is independent of wavenumber, so the depth of penetration for homogeneous surfaces with no concentration gradient was used. As indicated in the previous chapters, to account for these effects, the approaches in Chapter 5 should be used.

Polarization–Modulation Internal Reflection Experiment

In a search for more surface sensitivity and selectivity in IR analysis, Scanlon et al. (*48*) adopted their previously developed polarization–modulation method for IR reflection–absorption measurements (*49*) and illustrated that a similar experimental setup can be used in ATR measurements. Figure 9.12 is a schematic diagram of the instrumentation arrangements used for polarization–modulation ATR measurements. The basic design is the same as that used in reflection–absorption spectroscopy except that the reflecting sample is replaced by an ATR plate. The sensitivity is enhanced by a modulation of the polarized light, which allows p- and s-polarized light components with a modulation frequency ω to reach a detector alternately. The detector output is then routed through two separate channels. In one channel, the signal is demodulated using a lock-in amplifier tuned to a reference frequency ω_r. Since the output is proportional to the frequency at

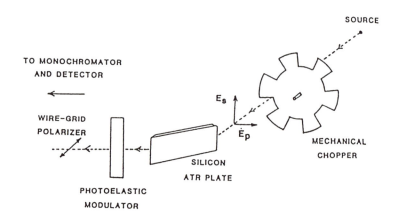

Figure 9.12. Schematic diagram of polarization–modulation internal reflection experiment. (Reproduced from reference 48, American Chemical Society.)

which the light is chopped (that is, the total radiation emitted from the source attenuated by the optics of the system), it is therefore proportional to the total radiation throughput of parallel and perpendicular polarized light intensities $I_\perp + I_\parallel$. The signal in the second channel is demodulated at 2ω, resulting in a lock-in amplifier output oscillating at ω_r, with an amplitude proportional to $I_\parallel - I_\perp$. The ratio $(I_\parallel - I_\perp)/(I_\parallel + I_\perp)$ gives the final output. The sensitivity enhancement is achieved by optical minimization of the bulk signal and enhancement of the intensity difference for p- and s-polarized light components. Although it seems that the use of modulation in ATR spectroscopy should be advantageous in monitoring early signs of surface treatments or modifications, only limited applications were reported (for example, an early detection of the SiOH species on a silicone surface).

Nano-Sampling Using ATR

As indicated earlier, ATR can be used on liquids and solids, and only minimal sample preparation is required. Under certain optical conditions, discussed in earlier chapters, interactions between electromagnetic radiation and the sample are stronger in an internal reflection mode than in transmission. Therefore, when light is localized and proper optical arrangements are used, it is possible to use ATR for micro- or nano-sampling applications. Harrick (50) offered a novel geometry, shown in Figure 9.13A, which consists of a light guide, an internal reflection prism, and a funnel. The light enters the prism through one of the end surfaces and travels down the

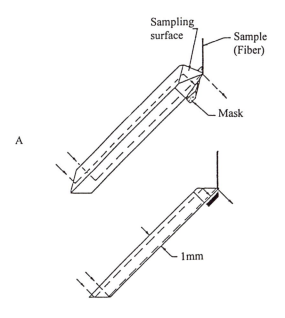

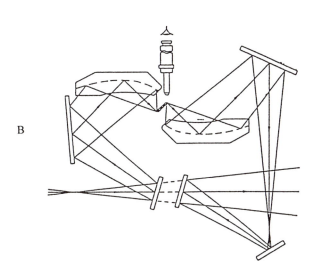

Figure 9.13. (A) Schematic diagram of a nano-sampling ATR accessory; (B) light path for a typical transfer optics for nano-sampling using ATR crystal. (Reproduced with permission from reference 49. Copyright 1987 Society for Applied Spectroscopy.)

crystal length until it hits a sample surface and exits the crystal. Liquid or solid sample is placed in contact with the crystal surface, and the positioning of a mask near an exit on the ATR crystal determines how much light is redirected to the detector. It appears that when the mask is not used, the sampling area is about 1 mm. Moving the mask closer to the edge of the crystal blocks a certain portion of the sampling surface. Using this approach, areas as small as 20 μm can be examined. Figure 9.13B shows the optical path of a commercially available transfer optics system for nano-sampling analysis by internal reflection.

References

1. Ishida, H.; Scott, C. *J. Polym. Eng.* **1986**, *6*, 201.

2. Snyder, R. W.; Fuerniss, S. J. *Appl. Spectrosc.* **1992**, *46*, 1113.

3. Barlick, E. G.; Messerschmidt, R. G. *Am. Lab.* **1984**, *16*(11), 18.

4. Bertie, J. E.; Eysel, H. H. *Appl. Spectrosc.* **1985**, *39*, 392.

5. Urban, M. W.; Koenig, J. L.; Shih, S. B.; Allaway, J. *Appl. Spectrosc.* **1987**, *41*, 590.

6. Urban, M. W.; Koenig, J. L. *Appl. Spectrosc.* **1987**, *41*, 1028.

7. Tiefenthaler, A. M.; Urban, M. W. *Appl. Spectrosc.* **1988**, *42*, 163.

8. Wong, J. S.; Rein, A. J.; Wilks, D.; Wilks, P. *Appl. Spectrosc.* **1984**, *38*, 32.

9. Bartick, E. G.; Messerschmidt, R. G. *Am. Lab* **1984**, *16*(11), 56.

10. Ishida, H.; Suzuki, Y. In *Composite Interfaces*; Ishida, H.; Koenig, J. L., Eds.; Elsevier: New York, 1986.

11. Mirabella, F. M., Jr., Ed. *Internal Reflection Spectroscopy: Theory and Applications*; Dekker: New York, 1993.

12. Tickanen, L. D.; Tejedor-Tejedor, M. I.; Anderson, M. A. *Langmuir* **1991**, *7*, 451.

13. Castillo, E. J.; Koenig, J. L.; Anderson, J. M.; Lo, J. *Biomaterials* **1984**, *5*, 319.

14. Brash, J. L.; Lyman, D. J. *J. Biomed. Mater. Res.* **1969**, *3*, 175.

15. Gendreau, R. M.; Leininger, R. I.; Winters, S.; Jakobsen, R. J. In *Biomaterials, Interfacial Phenomena, and Applications*; Cooper, S. L.; Peppas, N. A.; Hoffman, A. S.; Ratner, B. D., Eds.; Advances in Chemistry Series 199; American Chemical Society: Washington, DC, 1982; p 372.

16. Winters, S.; Gendreau, R. M.; Leininger, R. I.; Jakobsen, R. J. *Appl. Spectrosc.* **1982**, *36*, 404.

17. Jakobsen, R. J.; Brown, L. L.; Winters, S.; Gendreau, R. M. *J. Biomed. Res.* **1983**, *16*, 199.

18. De Lene Mirouze, F.; Boulou, J. C.; Dupuy, N.; Meurens, M.; Huvenne, J. P.; Legrand, P. *Appl. Spectrosc.* **1993**, *47*, 1187.

19. Urban, M. W. In *Polymers in Aqueous Media*; Glass, J. E., Ed.; Advances in Chemistry Series 223; American Chemical Society: Washington, DC, 1989; Chapter 15.

20. Wilks, P. A. In *Internal Reflection Spectroscopy: Theory and Applications*; Mirabella, F. M., Jr., Ed.; Dekker: New York, 1993.

21. Kuhn, A. T. *Techniques in Electrochemistry*; Wiley: New York, 1987.

22. Neff, H. *J. Electroanal. Chem.* **1983**, *150*, 513.

23. Neugebauer, H. *J. Electroanal. Chem.* **1981**, *122*, 381.

24. Pons, B. S. *J. Electron Spectrosc. Relat. Phenom.* **1987**, *45*, 303.

25. Tierney, M. J.; Martin, C. R. *J. Phys. Chem.* **1989**, *93*, 2878.

26. Bauhofer, J. In *Internal Reflection Spectroscopy: Theory and Applications*; Mirabella, F. M., Jr., Ed.; Dekker: New York, 1993.

27. Chabal, Y. J. *Surf. Sci. Rep.* **1988**, *8*, 211.

28. Muller, G. J.; Abraham-Fuchs, K. In *Internal Reflection Spectroscopy: Theory and Applications*; Mirabella, F. M., Jr., Ed.; Dekker: New York, 1993.

29. Neugebauer, H.; Neckel, A.; Nauer, G.; Brinda-Konopik, N.; Garnier, F.; Tourillon, G. *J. Chem. Colloq.* **1983**, *C10*, 517.

30. Pons, S. *J. Electroanal. Chem.* **1983**, *150*, 495.

31. Pons, S.; Davidson, T.; Bewick, A.; Schmidt, P. P. *J. Electroanal. Chem.* **1982**, *125*, 237.

32. Blajeni, B. A.; Habib, M. A.; Taniguchi, I.; Bockeris, J. O. *J. Electroanal. Chem.* **1983**, *157*, 399.

33. Bellissimo, J. A.; Cooper, S. L. *Trans. Am. Soc. Artif. Intern. Organs* **1984**, *30*, 359.

34. Pitt, W. G.; Cooper, S. L. *Biomaterials* **1986**, *7*, 340.

35. Kennedy, J. H.; Ishida, H.; Staikoff, L. S.; Lewis, C. W. *Biomater. Med. Devices Artif. Organs* **1978**, *6*, 215.

36. Ohnishi, T.; Ishitani, A.; Ishida, H.; Yamamoto, N.; Tsubomura, H. *J. Phys. Chem.* **1978**, *82*, 1989.

37. Pons, S.; Datta, M.; McAleer, J. F.; Hinman, A. S. *J. Electroanal. Chem.* **1982**, *160*, 237.

38. Simohony, S.; Katzir, A.; Kosower, E. M. *Anal. Chem.* **1988**, *60*, 1908.

39. Langmuir, I. *J. Am. Chem. Soc.* **1917**, *39*, 1848.

40. Blodgett, K. *J. Am. Chem. Soc.* **1935**, *57*, 1007.

41. Larsson, K.; Nordling, C.; Siegbahn, K.; Stenhagen, E. *Acta Chem. Scand.* **1966**, *20*, 2880.

42. Sugi, M. *Thin Solid Films* **1987**, *154*, 2163.

43. Takahara, A.; Morotomi, N.; Higashi, N.; Kunitake, T.; Kajiyama, T. *Macromolecules* **1989**, *22*, 617.

44. Chatzi, E. G.; Urban, M. W.; Ishida, H.; Koenig, J. L.; Laschewski, A.; Ringsdorf, H. *Langmuir* **1988**, *4*, 846.

45. Miyano, K.; Asano, K.; Shimomura, M. *Langmuir* **1991**, *7*, 444.

46. Comyn, J., Ed. *Polymer Compatibility*; Elsevier: New York, 1985.

47. Fieldson, G. T.; Barbari, T. A. *Polymer* **1993**, *24*, 1142.

48. Scanlon, K.; Kvitek, R. J.; Schulthesz, S. F.; Evans, J. F.; Overend, J. *J. Phys. Chem.* **1983**, *87*, 730.

49. Golden, W. G.; Dunn, D. S.; Overend, J. *J. Catal.* **1981**, *71*, 395.

50. Harrick, N. J. *Appl. Spectrosc.* **1987**, *41*, 1.

Appendix

ATR Configurations

Figure A.1 illustrates several light paths through various shapes of ATR crystals, and Table A.1 lists typical crystals used for internal reflection experiments. As shown in Figure A.1, crystal surfaces can be cut at different angles and in different shapes, making possible many geometric arrangements for various applications. Several configurations of the internal reflection attachments are shown in Figure A.2. Although the primary difference between each arrangement is how the incident light is directed through the crystal and redirected to the spectrometer, each configuration has its own attributes. For example, the multiple-reflection cylindrical crystal (D) is one of the best alternatives for studying aqueous solutions because water absorbs strongly in the IR region. Readers interested in the details of optical arrangements and designs should refer to Chapters 8 and 9 or more recent technical notes from the manufacturers. Note that there is a distinct difference between recording an ATR spectrum on dispersive and interferometric spectrometers. In principle, a single ATR cell cannot be used in both dispersive and interferometric spectrometers, because in the interferometric spectrometer the image of the beam is circular and often overfills a relatively small width of the ATR element edge, which in turn may cause all sorts of optical "leaks," leading to interference with the actual signal, wavenumber shifts, and band intensity distortions. Another important difference is the fixed or variable angle of incidence for cells. Although in many practical applications the angle of incidence is fixed at 45°, it is always desirable to be able to vary the angle because such an option allows surface depth profiling, one of the most attractive features of ATR spectroscopy.

Useful ATR Terms and Tips

Because measurements of ATR spectra involve reflections from the crystal–sample interface, it is appropriate to define several terms often used in spectroscopic jargon. The most common one, *plane specular reflection*, is the reflection from a polished flat surface, such as a common surface mirror shown in Figure A.3B. *Diffuse specular reflection*, or *scattered specular*

3348–9/96/0203/$15.00/0/© 1996 American Chemical Society

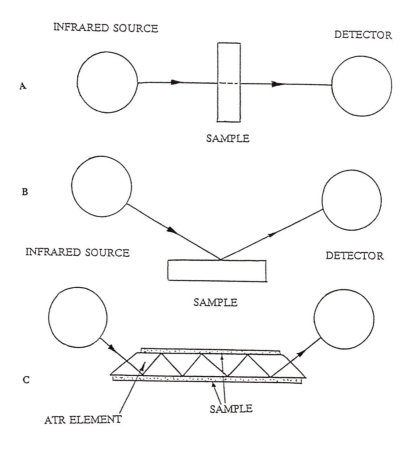

Figure A.1. (A) Transmission experiment: IR light passes through a sample, and a ratio of transmitted to incident light is measured. (B) Specular reflection experiment: IR light is reflected from a sample surface. (C) Attenuated total reflectance: light enters an ATR element and passes through, penetrating a sample–crystal interface.

reflectance, has nothing to do with diffusion and occurs when there are many single reflections in many directions. *Diffuse reflection* occurs when there are multiple reflections or transmissions each impinging on the surface ray within an IR-transparent sample (for example, KBr powder). The term $f/5$ is used to describe focusing and collection angle in IR spectrometers, and $f/1$ refers to the practical maximum (f is the focal length of the instrument). In practice, an $f/1$ or $f/5$ beam is introduced into the ATR crystal with a known refractive index. Usually, raising the refractive index of the crystal reduces the angular dispersion of the beam within the crystal.

Table A.1. Typical properties of selected optical materials used in internal reflection spectroscopy

Material	Useful range (μm)	Mean refractive index	Critical angle (°)	Property
Silver chloride	0.4–20	2.0	30	Soft, easily scratched
Silver bromide	0.45–30	2.2	27	Similar to AgCl, a bit harder
Zinc sulfide	5–16	2.2	27	Relatively hard and inert, water-insoluble, can be damaged by concentrated acids
Diamond	1–3.8 5.9–100	2.4	25	Hard
KRS-5 (thallous bromide iodine)	0.6–40	2.4	24.6	Favorable combination of properties
Zinc selenide	0.5–15	2.4	24.6	Expensive, water-insoluble, toxic when used with acids
Cadmium telluride	1.0–23	2.64	22.25	Expensive, relatively inert, can be used in aqueous studies
Arsenic selenide	1.0–12.5	2.8	20.9	Brittle, attacked by alkali solutions, some acids
Silicon	1.1–6.5	3.5	15.6	Hard, high resistivity, high-temperature applications
Silicon oxide (quartz)	0.3–2.3	1.43		Inert, desirable for low-index species in visible and ultraviolet
Gemanium	2.0–12	4.0	14.5	Good for fine depths of penetrations, temperature-sensitive (becomes opaque at 125 °C)

Reflection at a high angle of incidence increases penetration, but the choice of an angle near the critical angle may produce spectra with serious band distortions. In choosing the angle of incidence, first of all, the angle of incidence should not be too close to the critical angle. This estimation may be accomplished by comparing the $\sin^2\alpha_2$ term with the n_{21} values. This relationship should be such that $n_{21} < \sin^2\alpha_2$.

ATR experiments, as well as other spectroscopic techniques, require not only theoretical foundations for whose principles are addressed in the text, but also a fair amount of experience, which is briefly summarized

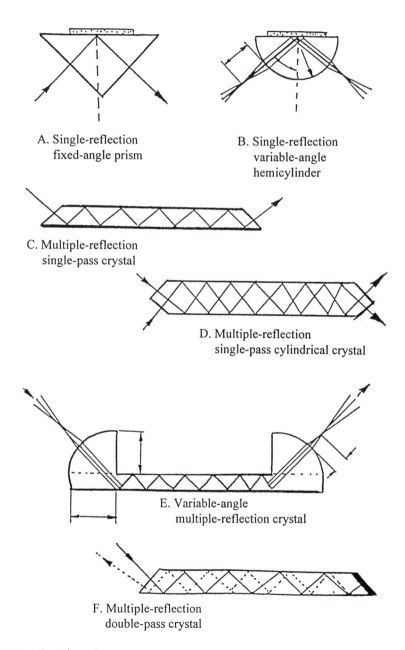

Figure A.2. Selected crystal configurations commonly used in ATR experimental setups: (A) single-reflection fixed-angle prism, (B) single-reflection variable-angle hemicylinder, (C) multiple-reflection single-pass crystal, (D) multiple-reflection single-pass cylindrical crystal, (E) variable-angle multiple-reflection crystal, and (F) multiple-reflection double-pass crystal.

A

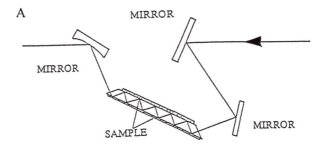

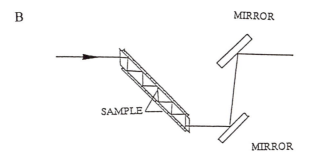

B

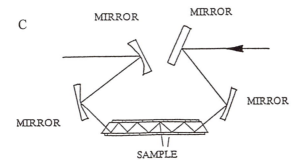

C

Figure A.3. Examples of ATR experimental arrangements.

here. In the first step, even before an ATR cell is placed in a spectrometer, it is common practice to check the alignment of the spectrometer. Then the ATR cell should be aligned in the spectrometer compartment, and a single-beam spectrum of an empty, properly aligned cell should be recorded. Such a single-beam reference spectrum is stored in the computer memory and represents the spectrum of the polychromatic source of radiation. Remember that the backing material used to attach a sample should not touch the ATR crystal; otherwise the baseline may drift. In the next step, the sample is placed on an ATR crystal and clamped tightly. Once all is placed in the ATR attachment, a single-beam sample spectrum is recorded and ratioed against the reference spectrum. Note that if the sample is a strongly absorbing species, it may be best not to cover the entire surface area of the ATR crystal. For example, silicone elastomers give strong absorptions because of the high dipole charges in the Si–O bonds. In this case, it is common practice to reduce the crystal coverage; for example, to attach the sample on only one side of the crystal. Once the cell is operational, the following practices are recommended:

1. ATR crystals should fit snugly in an ATR attachment.

2. Intimate contact between the sample and the crystal is a must and should be reproducible. Quan–Spec™ produces a cell that is equipped with a transducer for monitoring pressures applied to a sample–crystal interface.

3. ATR crystals should be free of scratches and should be as smooth as possible to maintain a good signal-to-noise ratio.

4. The crystal should be kept in good shape, but cleaned and handled as little as possible.

5. Uniform pressure should be maintained between the sample and the crystal as much as possible. A torque wrench is useful. Poly(tetrafluoroethylene) lining tape can be placed between the sample holder plate and the sample surface to ensure uniform pressure distribution.

6. No contaminants such as fingerprints, hand lotions, silicone lubricants, and sweat should be present.

7. If an ATR crystal is contaminated, it can be cleaned as follows:

   - Solid deposits can be blown away by clean compressed air or inert gas. (Warning: Compressed air may contain oil particles.)

   - Cast deposits can be removed by appropriate solvents (recommended are toluene, methylene chloride, acetone, and petroleum ether).

   - Fingerprints can be removed by using a water-soaked rayon ball and flushing with acetone. If a crystal is water-soluble, keep in mind that this treatment will dissolve the first few layers of the surface, which may be ideal, as long as the process results in a smooth, uniform surface.

   - Residual traces of impurities can be removed by ultrasonic cleaning. This is usually accomplished by hanging an ATR crystal in a beaker with a high-purity solvent and sonicating it in an ultrasonic cleaner filled with water. According to experienced experimenters in ATR, in a first treatment, you should use acetone as a solvent and sonicate no longer then 15 sec. In a second treatment, use high-purity petroleum ether as a solvent for 15 sec. (Warning: Longer treatments may destroy the crystal.)

- If a crystal has visible surface scratches, it should be repolished. It is advisable to send it to the manufacturer for repolishing.

Index

Bestsellers from ACS Books

The ACS Style Guide: A Manual for Authors and Editors
Edited by Janet S. Dodd
264 pp; clothbound ISBN 0–8412–0917–0; paperback ISBN 0–8412–0943–X

Understanding Chemical Patents: A Guide for the Inventor
By John T. Maynard and Howard M. Peters
184 pp; clothbound ISBN 0–8412–1997–4; paperback ISBN 0–8412–1998–2

Chemical Activities (student and teacher editions)
By Christie L. Borgford and Lee R. Summerlin
330 pp; spiralbound ISBN 0–8412–1417–4; teacher ed. ISBN 0–8412–1416–6

Chemical Demonstrations: A Sourcebook for Teachers,
Volumes 1 and 2, Second Edition
Volume 1 by Lee R. Summerlin and James L. Ealy, Jr.;
Vol. 1, 198 pp; spiralbound ISBN 0–8412–1481–6;
Volume 2 by Lee R. Summerlin, Christie L. Borgford, and Julie B. Ealy
Vol. 2, 234 pp; spiralbound ISBN 0–8412–1535–9

Chemistry and Crime: From Sherlock Holmes to Today's Courtroom
Edited by Samuel M. Gerber
135 pp; clothbound ISBN 0–8412–0784–4; paperback ISBN 0–8412–0785–2

Writing the Laboratory Notebook
By Howard M. Kanare
145 pp; clothbound ISBN 0–8412–0906–5; paperback ISBN 0–8412–0933–2

Developing a Chemical Hygiene Plan
By Jay A. Young, Warren K. Kingsley, and George H. Wahl, Jr.
paperback ISBN 0–8412–1876–5

Introduction to Microwave Sample Preparation: Theory and Practice
Edited by H. M. Kingston and Lois B. Jassie
263 pp; clothbound ISBN 0–8412–1450–6

Principles of Environmental Sampling
Edited by Lawrence H. Keith
ACS Professional Reference Book; 458 pp;
clothbound ISBN 0–8412–1173–6; paperback ISBN 0–8412–1437–9

Biotechnology and Materials Science: Chemistry for the Future
Edited by Mary L. Good (Jacqueline K. Barton, Associate Editor)
135 pp; clothbound ISBN 0–8412–1472–7; paperback ISBN 0–8412–1473–5

For further information and a free catalog of ACS books, contact:
American Chemical Society
Customer Service & Sales
1155 16th Street, NW, Washington, DC 20036
Telephone 800–227–5558